C000130749

THE CHEMISTRY AND PHYSICS OF COATINGS

Royal Society of Chemistry Paperbacks

Royal Society of Chemistry Paperbacks are a series of inexpensive texts suitable for teachers and students and give a clear, readable introduction to selected topics in chemistry. They should also appeal to the general chemist. For further information on selected titles contact:

Sales and Promotion Department
The Royal Society of Chemistry
Thomas Graham House
The Science Park
Milton Road
Cambridge CB4 4WF

Titles Available

Water *by Felix Franks*
Analysis—What Analytical Chemists Do *by Julian Tyson*
Basic Principles of Colloid Science *by D. H. Everett*
Food—The Chemistry of Its Components (Second Edition) *by T. P. Coultate*
The Chemistry of Polymers *by J. W. Nicholson*
Vitamin C—Its Chemistry and Biochemistry *by M. B. Davies, J. Austin, and D. A. Partridge*
The Chemistry and Physics of Coatings *edited by A. R. Marrion*

How to Obtain RSC Paperbacks

Existing titles may be obtained from the address below. Future titles may be obtained immediately on publication by placing a standing order for RSC Paperbacks. All orders should be addressed to:

The Royal Society of Chemistry
Turpin Distribution Services Limited
Blackhorse Road
Letchworth
Herts SG6 1HN, UK

Telephone: +44 (0) 462 672555
Fax: +44 (0) 462 480947

Royal Society of Chemistry Paperbacks

THE CHEMISTRY AND PHYSICS OF COATINGS

Edited by

ALASTAIR MARRION

Courtaulds Coatings (Holdings) Ltd.,
Gateshead, Tyne and Wear, NE10 0JY

A catalogue record for this book is available from the British Library
ISBN 0-85186-994-7

© The Royal Society of Chemistry, 1994

All Rights Reserved
No part of this book may be reproduced or transmitted in any form or by any means—graphic, electronic, including photocopying, recording, taping, or information storage and retrieval systems—without written permission from the Royal Society of Chemistry

Published by The Royal Society of Chemistry, Thomas Graham House, The Science Park, Cambridge CB4 4WF

Typeset by Vision Typesetting, Manchester
Printed and bound in Great Britain by
Redwood Books Ltd., Trowbridge, Wiltshire

Preface

This book is intended to expose some of the fascinating science encountered in the Coatings Industry to an audience well versed in chemistry, but perhaps having only nodding acquaintance with polymer science.

Though painting is one of the oldest polymer-based activities, it is also at the forefront of technology, and is an indispensable part of many manufacturing and construction processes. My fellow contributors and I describe, in simple terms, some of the remarkable variety of chemistry and physics which is called on to support the constant innovation in the Industry, and touch on the commercial and legislative background, without the extensive treatment of paint formulations for specific markets which forms a major part of most coatings text books.

The book is broadly organised in three sections. The first deals with economic and environmental aspects of the Coatings Industry, the second with the physics, and the third with the chemistry of coatings. Since it may be difficult for the uninitiated reader to grasp the contents of one section without some knowledge of the phenomena and terminology described in the others, we seek his forbearance and provide a glossary. A second purpose of the glossary is to supply systematic equivalents for the trivial chemical names frequently used in the Industry.

My co-authors are technologists, managers, and consultants within the industry, and I am indebted to them for finding time in their busy programmes to prepare their contributions. Any merit the work may have is due to them, whilst discontinuities and omissions are mine. We are grateful to Courtaulds Coatings (Holdings) Ltd. for its support of the work.

A.R. Marrion

Preface

[illegible]

Contents

Chapter 1

Economics and the Environment: The Role of Coatings

A. MILNE

INTRODUCTION

The principal concerns of our confused world are still the fundamental ones of food, shelter, and warmth, aggravated by an awareness of a growing human population, its economic and material well-being, and even survival, set against the finite resources of non-renewable fossil fuels and minerals, and the unequal access to them. World population is about 5.5 billion and growing at about 2% annum^{-1}, giving a doubling period of 35 years. Global energy production is about 8000 million tonnes oil equivalent (TOE), which is an average of 1.5 t person^{-1} highly unevenly distributed.

The Times Atlas gives the energy distribution shown in Table 1.1, within which there are even greater extremes. Norwegians, for example, are amongst the 8 t consumers, but they also export a further 16 t head^{-1}, and import an equivalent value in consumer goods.

As a simple comparison, a car-owner doing 15 000 miles annum^{-1} at 30 miles gallon^{-1} (Imp) is using rather more than 0.5 t of crude oil on motoring alone.

There is a close relationship between energy consumption, gross

Table 1.1 *Distribution of per capita Global Energy Consumption*

Population %	*Tonnes Oil Equivalent*
6.3	>8
21.3	2–8
64.3	0.2–2
7.8	<0.2

domestic product, and standards of living. There is a remarkable consistency, for example, in the paint and coatings consumption of the developed world (Table 1.2). It is difficult, however, to see how the aspirations of two-thirds of the doubled world population, consuming less than 2 TOE can be satisfied, as well as satisfying the other demands of energy and resource conservation, acid rain reduction, and global warming.

Table 1.2 *Global paint market*

Territory	*Production* $/10^6$ t	*Population* $/10^6$	*Per capita* /Kg year^{-1}
USA	4.9	250	20
EEC	5.3	340	15.6
Former USSR	3.3	220	15
Japan	1.9	120	15.8

What has all this to do with coatings? A great deal. Somewhat serendipitously, but from historically very sound roots, the coatings industry finds itself an inconspicuous but significant contributor to the solution to some of the problems. We are able to say, with a degree of complacency, that we have always been in the business of conservation as well as decoration. We are not, however, allowed that degree of complacency. Seen from the outside as major users of heavy and toxic metals, volatile hydrocarbons, halogenated monomers and polymers, highly reactive chemicals, and carcinogens, we are perceived as part of the problem rather than as part of the solution. Part of the purpose of this book will be to see how the industry can rise to the challenge of changing priorities and demands being put upon it. Increasingly, environmental concern is followed by legislative constraints. The original concerns of the environmental movement, for example, were with the basic extractive industries, wood, water, coal, iron, and agricultural products, and more recently with oil, gas, uranium, and renewables and non-renewables of all kinds. What follows in this book is an outline of some of the ways in which the industry is responding, and can respond, to the changing demands being put upon it. For the more fortunate 20% of the world population, for example, with a disposable energy budget of 5–8 TOE head^{-1}, it is not difficult to make alternative choices and there is an increasing preference for, and an increasing value being put upon, clean air, clean water, natural beauty, and an increasing willingness to pay for them. No doubt these would be preferences of the other 80% also!

The most economic solution to most problems, following directly from the Second Law of Thermodynamics, is the reduction and elimination of pollution at source. Indeed, if we can express our problem in concentration terms we can get a first approximation to the cost, as in equation (1):

$$\Delta G = RT \ln C_1/C_0 \tag{1}$$

where ΔG is Gibbs free energy change, R is the Boltzmann constant, T is absolute temperature, and C_1 and C_0 are the starting and finishing concentrations, always remembering that in the real world, efficiencies of real reactions are rarely more than 20%. That gives us a molar cost for pollutant and effluent recovery. Hence the emphasis on total elimination at source, and the emphasis in this book of strategies for avoidance of problems. Forethought would seem to be one of the few thermodynamically efficient processes! If the objective is energy conservation then recovering dilute liquid and gaseous pollutants very quickly becomes self-defeating in energy terms.

In the following chapters therefore, the authors respond to demands to remove solvents, by formulating coatings on liquid oligomers, use little or no solvent, as powder coatings without solvent, use liquid only at elevated temperature for a few minutes, or in the form of aqueous latices or non-aqueous dispersions. There is an emphasis on doing more with less by radical improvements to the performance of coatings. Each new technique and material creates its own problems and opportunities. Highly reactive monomers and oligomers are more likely to be hazardous almost by definition, and water, from an environmental viewpoint, would seem to be an ideal dispersant in everything except boiling point and latent heat of evaporation, while under ambient conditions it brings its own problems at low temperature and high humidity.

FUNCTIONS OF COATINGS

Why coatings? Why not simply paints? Historically there was no distinction. Leonardo da Vinci used essentially the same materials and methods of preparation and application as the house painters of his day, and the present professional association in the UK, 'The Oil and Colour Chemists Association', is only now seeking a title more in keeping with the present role of its members. When we make the distinction between paint and coatings we imply a distinction between functions which are largely decorative and aesthetic in the former, and have a more serious purpose in the latter. Indeed there are several serious purposes, and they would usually be expressed in specific terms. Some of the specific purposes are:

the prevention of corrosion, either actively, by the inclusion of anticorrosive pigments, or passively by providing an adhesive and impermeable barrier; providing either slip or slip resistance; impact and abrasion resistance; resistance to contamination; or hygienic properties, such as fungal, bacterial, or fouling resistance. Whatever the desired end property, what they have in common can in general be expressed in economic terms and calculated in terms of:

(i) energy savings,
(ii) reduced down-time,
(iii) increased life-time,
(iv) capital savings,
(v) materials substitution.

The distinction between paint and coatings, even in economic terms is more apparent than real, since even aesthetic choices can be accommodated in standard economic theory in terms of willingness to pay for some benefit, or willingness to pay for the elimination of some undesired consequence, the 'Hedonistic Price Principle'.

Easiest to calculate are the immediate and direct benefits, the producer surplus and the consumer surplus, since they are the everyday transactions of supplier and consumer. We can illustrate (i)–(iv) above for the case of antifouling use, for example, since this is one of the few areas where the benefits are most readily calculated, in energy and monetary terms, and since all antifoulings (to date) have used biocides which come under specific legislation, to which both suppliers and users have had to accommodate. Antifouling coatings are thus a paradigm for many of the constraints being imposed by increasingly stringent legislation, one of which is the possibility that when other and incompatible preferences are being expressed, the solution is sub-optimum in the very area for which the coating was designed, *i.e.* maximum energy conservation at least cost. As coatings, antifoulings also illustrate at least two of the less expected functions which can be built into coating materials. Permanence is usually regarded as a desideratum in a coating; self-polishing antifoulings are designed to disappear, and smoothness, usually an aesthetic consideration in a coating, is here intimately connected to the turbulent flow over the surface, again with desirable energy-saving consequences. Marine fouling and hull roughness increase the drag on vessels, and that increase can be readily translated into fuel consumption. The world fleet of some 39 000 vessels burns 184 million t of heavy fuel oil, at a cost of $100 (US) t^{-1}, or $18.4 billion. The base line, the cost of doing nothing, would mean that the whole fleet would burn 40% more fuel, an increase of 72 million t. In practice since most vessels operate close to maximum power, the fleet would have to increase by a similar amount. For comparison, the UK sector of the North Sea produced at maximum, just over 100 million t of oil annum^{-1}.

The increase in the price of oil, in 1973, from about £5.00 t^{-1} to £50.00, and in 1979 to £100 t^{-1}, required a radical improvement in antifouling performance and in antifouling lifetime. The improvement took the form of the self-polishing tri-butyl tin copolymer antifoulings. Improvements in fouling protection, and in ship roughness, save 4% of fuel, 7.2 million t, valued at $720 million.

At the same time ship-owners demanded much longer lifetimes between drydocking. Traditionally vessels had docked on an annual basis. This was increased to 30 months and the current maximum is 5–7 years. The average improvement in 10 years (1976–86) was from 21 to 28 months. This has a value of $820 million. Since average docking was for about 9 days, two days annum^{-1} extra trading was also available, worth $420 million. Capital savings are more difficult to calculate but a typical 200 000 DWT tanker will carry about 1 million t annum^{-1}, so the above 7.2 million t oil saving is, itself worth 7 tankers at $80 million each, worth in annual opportunity cost some $500 million. So, for a modest improvement in antifouling performance, a total saving of about $2.5 billion is obtained. This is obtained also at no nett increase in price of antifoulings. The global antifouling market is about 25 million l, at $8 l^{-1}, so the $2.5 billion benefit is purchased for approximately $200 million, a benefit-cost ratio of 11.5. Both producers (the coatings manufacturers) and the consumers (the ship-owners and operators) are happy. In the language of cost-benefit analysis there is clearly a producer and consumer benefit. Resources are saved (oil and steel to make seven vessels), and those very environmental aspects on which such weight is placed, acid rain and the alleged 'greenhouse effect', benefit to the extent of 560 000 t and 23 million t respectively. But the producer and consumer are no longer the only participants in the transaction. Tri-butyl tin is a 'Red List' pesticide, and it has measurable environmental effects both on economic crops (oysters) and non-economic species such as the dog-whelk, *Nucella lapillus*. No one, it is thought, has yet evaluated these environmental impacts, but the case serves to illustrate some of the ramifications of environmental risk-benefit analysis. For example, using the 'Hedonistic Price Method', one would be required to put a value on the pleasure of finding dog-whelks on a particular stretch of coast, or to put a value on bio-diversity. Exactly how much pleasure, to how many people, for what period is discussed in the literature. Alternative methods available are the 'Contingent Valuation Method', which claims to be able to put a value on the existence of a resource whether one uses it or not, or might wish to in the future, and the 'Replacement Cost Method', which attempts to evaluate the cost of recreating a damaged habitat or ecosystem. Consideration of human safety and health require one to estimate a value-of-statistical-life. This is commonly done in the transport field where the values of investment in accident prevention have to be calculated, but the methods apply equally in the fields of industrial accidents and the handling and use of hazardous materials.

This would seem to take us into territory remote from the design of coatings fit for a variety of purposes, but the process has been considered for one of the many raw materials in the coatings field, and will be so increasingly in other areas. Such considerations are clearly the preoccupations of those 20% of the population burning 8 t of oil equivalent; they are unlikely to feature strongly in the minds of those burning less than 2 t.

Item (v) in the above list of criteria, materials substitution, is by far the most important aspect of the 'added value' of coatings. There is an inverse relationship between the value and abundance of metals in the Earth's crust. Platinum, gold, silver, copper, and tin, are rare, precious, and coinage metals, comprising no more than 71 $g\,t^{-1}$ of the Earth's crust, while iron and aluminium at 5.8% and 8.8% of the Earth's crust respectively are by far the most abundant. Coated steel performs many of the functions which would otherwise require much more expensive metals and alloys, such as copper, bronze, or stainless steel or much more expensive coating processes such as ceramic coating, electroplating with expensive metals such as nickel or chromium, vacuum metallising, or plasma coating. Coatings therefore fulfil some of the ambitions of the alchemists, in transmuting some of the properties of base metals into gold. This is 'added value'. Global production of iron and steel is about 700 million t $annum^{-1}$, or about 9 t $person^{-1}$ $(average\ lifetime)^{-1}$ of 70 years. Iron and aluminium occur in large deposits, in high concentration at or near the Earth's surface. The major component of the cost is the energy required to reduce them to the metallic state. They are thermodynamically unstable in our oxidising atmosphere, and tend to revert to the original oxides, as corrosion products. The materials thus substituted at great energy cost, require to be conserved.

We are also highly uneven in our admiration of corrosion products. We admire the patina on copper and bronze, but have only a limited admiration for galvanised iron, and none at all for rust. There was an attempt some years ago amongst new-brutalist architects to get us to like rust, and the metallurgists complied by producing a uniformly rusting steel, Cor-Ten® steel; but we refused to love it. So one of our primary coatings objectives has been, since the Bronze Age, the substitution of semi-precious metals by iron and steel. We can get some idea of how much this is worth to the users in the form of an approximate cost-benefit.

At the stage of iron ore, about 50% Fe, the cost of iron is about £13 t^{-1}. The energy requirement is 60–360 GJ t^{-1}, and at a cost of 2.2 p (kW h^{-1}), or an oil price of £67 t^{-1}, the energy content of a tonne of steel is about £100, *i.e.* most of the cost is 'pure' energy. The added value of a coating is difficult to calculate, depending both on the gauge of the steel, the thickness of the coating and the end use. For some applications the uncoated steel would be of zero value even for the shortest period of time, *e.g.* a beer can uncoated inside or outside, but for a long term use, a bridge or a ship say, calculation is possible.

An uncoated ship, in sea-water would suffer severe corrosion in five years or less. An adequately coated one might avoid serious repair for 15–25 years. A super-tanker might cost $80 million. If, by coating the lifetime is tripled, then the saving is $160 million on 16 000 t of steel, or $10 000 t^{-1}. The coating enhances the value by fifty times. If the steel plate is 15 mm thick, and protected on both sides with 250 μm of a polymer coating, then 120 kg (1 m^2), of fabricated steel, worth $1250, is protected for 25 years for about $2 of polymer, a benefit-cost ratio of 600 times.

In those areas where coated steel replaces, say, stainless steel, similar benefits are achievable. Indeed, in the area of chemical resistance, coated mild steel is suitable for most likely cargoes, with stainless steel used for only the most aggressive chemicals. In the case of both steel and aluminium, the energy required for remelting is a small fraction of that required for the original reduction. Coating, in both cases, ensures that almost the total original weight of metal is available for recycling. These two illustrations of direct energy saving and resource conservation are the principal contributions of coatings to the community.

Chapter 2

Volatile Organics—Legislation and The Drive to Compliance

D.J. WIGGLESWORTH

ATMOSPHERIC POLLUTION AND THE COATINGS INDUSTRY

Volatile Organic Compounds (VOCs) are substances which contain carbon and which evaporate readily, though a few of these materials such as the oxides of carbon are not generally classed as VOCs. Most solvents used in paint, perfumes, adhesives, inks, aftershave, *etc.* are VOCs and other VOCs are present in various household chemicals, petrol, dry cleaning fluids, car exhaust fumes, cigarette, and bonfire smoke, *etc.* Within Europe some 40% of all VOC emissions to the atmosphere arise from natural sources such as that invigorating smell of pine needles in some forests, emissions from animals and agriculture in general, *etc.* Some estimates of total world VOC emissions put the natural contribution at around 90%. Most of the adverse effects of VOC emissions arise because of the type of VOC, the local concentration, or the combination with other air pollution, and because of the concentration effect much of the difficulty caused by VOCs arises from anthropogenic releases.

Carbon dioxide, nitrogen oxides (NO_X), and sulfur dioxide are not VOCs but are also emitted to the atmosphere and, either alone or in association with VOCs, contribute to a range of current or potential atmospheric pollution effects, the most well known examples of which are reviewed below.

The Hole in the Ozone Layer

Some 15 to 30 miles above the Earth, in that level of the atmosphere called the stratosphere, there is a layer in which the concentration of ozone is quite high. This occurs above the first low temperature trough of the atmosphere and so there is very limited mixing of gases back to the lower level of the atmosphere (the troposphere).

Some VOCs are long lived in the lower atmosphere and slowly migrate to the stratosphere, and once there can cause particular effects. One such group is the family of chlorofluorocarbons (CFCs). These are used as coolant liquids in refrigerators, deep freezers, and air conditioners, *etc*. In the stratosphere these compounds are photolytically dissociated by incoming sunlight to form chlorine oxide and chlorine which reacts with the ozone to form oxygen and also regenerates a high proportion of the chlorine oxide. Thus the process continues and the ozone layer is depleted (Scheme 1).

$$\text{CFCs} \xrightarrow{h\nu} Cl_2$$
$$Cl_2 \xrightarrow{h\nu} Cl^{\bullet} + Cl^{\bullet}$$
$$2O_3 + 2Cl^{\bullet} \longrightarrow 3O_2 + Cl_2$$
$$O_2 \xrightarrow{h\nu} O^{\bullet} + O^{\bullet}$$
$$2Cl^{\bullet} + O_2 \longrightarrow 2ClO$$
$$ClO + O^{\bullet} \longrightarrow Cl^{\bullet} + O_2$$

Scheme 1

In some areas of the world, and particularly in the Antarctic, the level of depletion is quite marked and hence the expression 'hole in the ozone layer'. There are undoubtedly other effects at play as well which probably include the influence of 'solar winds' and of volcanic eruptions, and both polar stratospheric clouds and sulfur compound particles emitted by volcanoes can provide surfaces for the catalytic formation of active chlorine. Nonetheless there is now wide acceptance that most of the chlorine comes from CFCs and that there is therefore a direct link between CFCs and stratospheric ozone depletion. The lack of ozone at this level increases the transmission of the more harmful parts of the sun's radiation and is expected to result in increased levels of skin cancer and eye cataracts. Most of the world's developed nations have agreed to phase out the production and use of the CFCs during this decade under the so-called 'Montreal Protocol' but the releases may continue from existing equipment and the problem will remain for very many years at least.

Summer Smogs

Meanwhile at lower levels, in the troposphere, there is a different effect which is that of ozone creation. At this level VOC emissions react under the influence of sunlight and in the presence of nitrogen oxides to convert oxygen into ozone. This is one part of the process that generates summer hazes and smogs which are frequently seen and become quite severe in cities such as Los Angeles and Athens among many others. Essentially what is happening is that under the effect of strong sunlight VOCs photolytically dissociate and then react with lower oxides of nitrogen to yield aldehydes and higher oxides of nitrogen. Further sunlight subsequently causes the reformation of the lower oxides of nitrogen and an excess of ozone. The process only requires continuing sunshine and more VOC to continue on and on (Scheme 2). The rate of reaction coupled with wind speeds and potential delayed reaction due to low intensity of sunlight can mean that the ozone and the resulting smog can be formed at substantial distances from the original emission source.

$$O_3 + H_2O \xrightarrow{h\nu} 2\,{}^{\bullet}OH + O_2$$
$${}^{\bullet}OH + RH \xrightarrow{O_2} RO_2^{\bullet} + H_2O$$
$$RO_2^{\bullet} + 2NO \xrightarrow{O_2} R^1CHO + 2NO_2 + {}^{\bullet}OH$$
$$NO_2 + O_2 \xrightarrow{h\nu} NO + O_3$$

Scheme 2

The resulting smog can cause eye irritation and breathing difficulties and is generally unpleasant. However, ozone is also toxic and in many areas of the world, summer time excursion levels are at or above the levels rated as harmful to crops and trees and at least irritating to animals, including us. The World Health Organisation guideline levels are 8 hour average of 50–60 p.p.b. and yet higher levels are frequently recorded in the UK and Continental Europe. Many countries have legislated, or are now legislating, to reduce both VOC and nitrogen oxides emissions as part of the overall effort to reduce tropospheric ozone creation but the political issues and the chemistry are both very complex. Emissions in one state can give rise to pollution several days later in another country hundreds of miles away and with this as background The United Nations has sponsored a Protocol calling for a reduction of 30% in emissions of VOCs by the end of this decade. Most European countries and the USA and Canada have signed the Protocol but the issue is how to achieve the necessary reduction in ways which are politically and economically acceptable. The problem remains and will get worse before it gets better. The stability of some VOCs and their relative decomposition rates under various reaction conditions greatly complicate the issues.

Other Health Effects

Apart from health effects of ozone creation, most VOCs have some toxic effects themselves and some are very toxic whilst others are carcinogens. This problem is mainly solved in most developed countries by careful controls within industrial plants and the reduced use of household coal fires and steam trains. These effects are classified as local issues by most nations. The VOCs from smoking tobacco continue to affect some.

The Greenhouse Effect

There have always been variations in the Earth's climate and substantial data show that this has been happening for centuries. However there is growing concern that pollution may be causing a shift that will not be part of this natural variation and will not be naturally reversible.

The greatest level of concern is over the 'greenhouse effect' and its influence on global warming. World knowledge is insufficient to know the extent to which this might happen or the total effect if it did. Nonetheless, the principle is well understood and supported and the potential problem is under close scrutiny.

The effect is that certain atmospheric gases, such as carbon dioxide, several VOCs, ozone, CFCs, and others, will transmit a high proportion of sunshine but transmit less of the long wave radiation emitted by the Earth (just like the effect of glass in a greenhouse). So there is a net increase in the temperature of the Earth. Of course this has been happening forever and some estimates are that the Earth would on average be 30°C cooler if the 'Greenhouse Effect' did not exist.

However, the issue at stake is that the concentration of these gases in the atmosphere is increasing, and it is uncertain what the effect of this will be. In theory as the gas concentration increases so the energy retained by the Earth should rise and further global warming should take place. Hence the efforts to reduce the emissions of these gases through controls and possible carbon taxes, *etc*. This becomes an issue for the paint industry partly due to the implied need to use extra energy if incineration is used to abate VOC emissions.

Acid Rain

Acid rain is mainly caused by sulfur dioxide and nitrogen oxides emissions returning to Earth as the respective derived acids. However, some VOCs, especially CFCs, can also contribute to this. A small proportion of emitted CFCs return to Earth as hydrochloric acid in rainfall.

The most significant effects of acid rain are damage to vegetation and to aquatic life in freshwater. Legislative focus is on reducing emissions of sulfur dioxide from power stations and nitrogen oxides from internal combustion engines and the degradation of man-made fertilisers.

So these are the major concerns, they are complex issues and this is reflected in the resulting regulations.

LEGISLATIVE REACTIONS

Many legislators around the world have been tackling the issues raised by VOC emissions as part of their overall concern with air quality and whilst it is beyond the scope of this book to examine all of these actions it is nonetheless necessary to review some regulations in order to understand the part that legislation and legislative proposals play in influencing the strategies of the coatings industry.

Through the 1970s and 1980s some major issues of air pollution have been increasingly recognised as world problems requiring global action to achieve significant benefit. This realisation has resulted in the United Nations taking an increased part in achieving Conventions which start to address these issues. Various agencies of the United Nations Organisation are involved, *e.g.* United Nations Environment Programme (UNEP), United Nations Economic Commission for Europe (UNECE), United Nations Conference on Environment and Development (UNCED), *etc.*

These bodies provide a forum for discussion and co-operation, make strong recommendations, and publish detailed Convention statements and Protocols. However, ultimately it is for national governments to introduce and enforce the implementing legislation, or not, as can be the case.

Possibly the best known action is that to protect the ozone layer. In response to concerns that had been growing for some years UNEP held a series of meetings that generated an 'Agreement on Substances which Deplete the Ozone Layer'. This was achieved in Montreal in September, 1987 and became known as the Montreal Protocol. The Protocol was ultimately signed by over 60 governments who made commitments to reduce consumption of CFCs by 50% by 1999 and to freeze production of Halons at 1992 levels. Subsequent meetings have been held to review depletion data and causative evidence which have encouraged a commitment to even stronger action. Further work will probably serve to accelerate even more both the scope and rate of these actions. Proposals already foresee a total ban on production of CFCs, Halons, 1,1,1-trichloroethane, and carbon tetrachloride and place the future of hydrofluorocarbons (HCFCs) formally on the agenda for subsequent meetings. All of this has impacted on the options that the paint industry has for fighting fires (Halons), for formulating aerosols (CFCs), for manufacturing chlorinated rubber (carbon tetrachloride), and will probably have some bearing on the use of some low POCP (Photochemical Ozone Creation Potential) solvents (1,1,1-trichloroethane) and the formulation of paint strippers (methylene chloride).

Another area which has attracted United Nations attention has been tropospheric air quality. UNECE developed a forum to discuss air quality which resulted in 35 parties, mainly national governments, ratifying or accepting a Convention on Long-Range Transboundary Air Pollution which was adopted in Geneva on 13 November, 1979.

This Convention has spawned four Protocols to date. The first three covered:

- Financing the Monitoring and Evaluation of Long-Range Transmissions of Air Pollutants—adopted 28 September, 1984.
- Reducing Emissions of Sulfur—adopted 8 July, 1985.
- Reducing Emissions of NO_X—adopted 31 October, 1988.

Whilst these are not directly associated with VOCs it is, as noted previously, well established that it is the combination of VOC and NO_X concentrations in the atmosphere that are the major cause of ozone creation.

The fourth Protocol addresses the issue of VOC emissions directly. This is the 'Protocol to the 1979 Convention on Long-Range Transboundary Air Pollution concerning the Emissions of Volatile Organic Compounds or their Transboundary Fluxes' which was adopted 18 November, 1991. By the end of 1992 the Protocol had been adopted by 23 parties, mostly national governments, and ratification or acceptance is expected through 1993/4.

The VOC Protocol is not addressed directly to industry but relies on national legislation to give legal effect to its proposals. The Protocol allows for various compliance standards and most signatories have selected the option to reduce their VOC emissions by at least 30% by 1999 using 1988 emission levels as the base-line. Some countries which felt that they had already made significant reductions before 1988 negotiated for an earlier base year, *e.g.* the USA has committed to the minimum 30% reduction by 1999 using a base-line of 1984, whilst some whose emissions were still increasing in 1988 chose 1990 as their base-line, *e.g.* Italy. Some countries which have relatively weak economies have committed only to freeze their emissions, *e.g.* Greece will freeze at 1988 levels. Canada and Norway have each chosen the option of declaring only part of their country as a Tropospheric Ozone Management Area (TOMA) and their commitment to emission reduction applies only within the TOMAs. All of these options are valid compliance methods but of course focus only on the desired result not on the means of achievement nor even on the share of the burden to be borne by different VOC emitters. The Protocol itself identifies four control options:

- substitution of VOCs by the use of alternative products which are low in or do not contain VOCs.
- reduction of VOC emissions by best management practices in housekeeping, maintenance, or processing.
- recycling and/or recovery for re-use of VOCs.
- destruction of potentially emitted VOCs by control techniques such as thermal or catalytic incineration or biological treatment.

The UNECE VOC Protocol classifies VOCs into three groups based on the degree to which they contribute to the formation of episodic ozone, with the aromatic hydrocarbons, which make up a high proportion of the solvents used by the paint industry, being in the most damaging category. The Protocol also provides some numerical quantification of this importance by introducing the concept of 'Photochemical Ozone Creation Potential' (POCP) and provides detailed tables of POCP values according to various test regimes. Broadly these reflect the reactivity of particular solvents within the relevant chemical mechanisms. Unfortunately many of the less 'reactive' solvents give rise to other problems, *e.g.* benzene and chlorinated hydrocarbons are both classified as 'least important' yet benzene is a carcinogen and chlorinated hydrocarbons can contribute to the depletion of the ozone layer.

Maybe the greatest contribution that the Protocol will make to these complex issues is that it requires the publication of a proposed compliance strategy by each signatory. Thus substantial international attention will be focused on these issues in the period 1993–95 and the proposals, which will all be in the public domain, are likely to generate new or more focused approaches.

The European Commission is drafting a directive to address solvent emissions and this derives in part from their commitments under the UNECE Protocol. The directive will be addressed to member states and currently has a structure of a series of articles, which establish the general direction, coupled with annexes which define methodology for Solvent Management Plans, provide guidance on the selection of solvents, and set specific requirements on emission limit values for each of many individual sectors of solvent use. These currently include some sectors not associated with the paint industry but do specifically include paint manufacturing processes and the usage of paint by several customer groups such as car manufacturers and the industrial coating of metal, wood, and plastics. This directive deals only with emissions from contained processes and a second directive is expected to follow and will address the issues arising from non-contained processes, *e.g.* the use of decorative paint.

In the meantime real control is still nationally based and within the national legislatures the introduction and growth of air quality legislation has come at different times and with varied drivers. In the UK the 1956 Clean Air Act derived directly from the successive London Smogs which culminated in the so-called 'Killer Smog' of December, 1952 and accordingly focused on sulfur oxides and particulate pollution. In the mid-1960s the 'Rule 66' and later 'Rule 102' of Los Angeles aimed to combat urban smog by discouraging the use of certain solvents which were particularly active in promoting the formation of ozone. This concept is now carried through to the UNECE VOC Protocol as a consideration of POCP values. In Germany the early introduction of 'TA-Luft', the Technical Instruction on Clean Air, addressed the issues of direct health effects as well as those of ozone creation.

Summarising each of the major pieces of national legislation is fraught with difficulties, can be misleading, and will always grossly oversimplify the issues and control parameters. The reader is therefore most strongly directed to the source documents in their then current form before drawing any specific conclusions on any nation's legislation. That said there are some interesting differences in approach and control detail between some of the current national requirements.

In the USA the Clean Air Act Amendments 1990 provided a substantial uplift to the existing law by adding over 700 pages of new and amended requirements, and set an agenda which will run until well into the 21st century. It addresses a very wide programme of air quality improvements and is structured around seven titles and a series of appendices.

Its major provisions affecting the paint industry are:

— to reduce VOC emissions as a contribution towards reducing ozone levels in 'ozone non-attainment areas'.

— to control emissions of 'Hazardous Air Pollutants' (HAPs) from sources with the potential to emit 10 tons per year (TPY) of any one HAP or 25 TPY of any combination of HAPs. The Act lists some 190 compounds to be subject to 'Maximum Achievable Control Technology' (MACT). This list includes several solvents which are of importance as solvents in the paint industry, *e.g.* methyl ethyl ketone, toluene, xylene, methyl isobutyl ketone, *etc.*

— mandatory operating permits, that will affect most significant paint manufacturers and paint users, with a guidance fee of \$25 per ton of regulated pollutant.

— elimination of substances that deplete the ozone layer in the stratosphere (fulfilling Montreal Protocol commitments).

— provisions relating to civil and criminal penalties for effective enforcement. The Act authorises fines on individuals up to \$250 000 and five years imprisonment.

Relative to the USA Clean Air Act 1990 most other national legislation is both short and less restrictive but the tendency is likely to be towards the USA model at least in the depth of detail though not necessarily in exactly the same direction.

In Germany, air quality, including emissions of VOCs, is controlled under 'TA-Luft', the Technical Instruction on Clean Air, which is an administrative regulation aimed at Local Government but indirectly controls industry via the process approval procedures of installations subject to license. TA-Luft distinguishes between carcinogenic and non-carcinogenic materials, and for each set of materials three classes are defined according to the assessed potential hazard. Maximum emission limits are specified for each class and these range from $0.1\,\mathrm{mg\,m^{-3}}$ for carcinogenic Class I to $150\,\mathrm{mg\,m^{-3}}$ for non-carcinogenic Class III material. For each class and sector mass flow rates are specified above which an installation comes in scope. There are two licensing categories based on weight of solvent used per hour and the more onerous control requirements apply to installations using more than $250\,\mathrm{kg\,h^{-1}}$. Special rules apply to some major industry sectors, for example the motor industry is controlled on weight emissions of non-carcinogenic VOC per square metre of car coated rather than by the general industrial requirement based on concentration of VOC in exhaust air.

In Italy emissions to atmosphere are controlled by two major laws. DPR203 defines those installations in scope and the authorisation procedures, whilst DM51 sets emission limit values for air pollutants of which five classes are specified with minimum flow rates above which an installation is controlled to emission limit values which range from $5\,\mathrm{mg\,m^{-3}}$ for Class I to $660\,\mathrm{mg\,m^{-3}}$ for Class V. There is a special requirement placed upon ovens of industrial coating plants not to exceed emissions of $50\,\mathrm{mg\,m^{-3}}$ measured as carbon. The introduction of DM51 is phased until 1997 with major polluters having earlier introduction dates.

In the Netherlands, large emitters of VOCs require an authorisation based on Best Available Technology (BAT) from local government under the Air Pollution Act of 1972 prior to the operation of such an installation. However all of this is largely overshadowed by a voluntary programme

KWS2000 between industry, government, and the people of Holland in which they aim to reduce VOC emissions by 50% by 2000 using 1981 as a base year. Regular reports on progress are published which make impressive reading and are available in English from the Dutch government.

In the UK the Environmental Protection Act 1990 places controls on the emissions from most contained use of paint under Part B of the Act. Many Guidance Notes have been produced to guide industry towards best compliance. In general releases of VOCs are to be controlled to less than $50\,\mathrm{mg\,m^{-3}}$ measured as carbon, though derogation is achieved if coatings of low VOC content are used. Extensive guidance tables are published which define VOC contents of each major coating in each sector, necessary to achieve compliance.

Until recently most of the legislation on emissions of VOCs to atmosphere have tended to focus on air quality issues. The World Commission on Environment and Development recognised the necessity for integration of, and interdependency in, environmental pollution control in their paper entitled 'The Institutional Gap'. It is now widely recognised that no single medium in the environment is independent of any other and that emissions to one should not be controlled without reference to the consequences to other media. Increasingly through the 1990s the legislators will take greater account of the overall environmental balance and require detailed assessment of effects on soil and water before authorising revised emissions to air. France has a long history of taking an integrated approach to pollution but in most countries the approach is quite new. The integrated Pollution Control sections of the UK Environment Protection Act 1990 for Part A processes is a good example of the recent application of the concept. The European Community is beginning to address the same issues and is drafting a directive on Integrated Pollution Prevention and Control (IPPC) which is likely to become the major framework directive on industrial environmental releases with directives on such issues as the reduction of solvent emissions being effectively daughter directives to the IPPC directive.

PAINT INDUSTRY RESPONSE AND STRATEGY

Paint is used by so many different market segments each with unique facilities, challenges, and requirements that to prevent emissions from all and from the paint manufacturing process requires the availability of a wide range of control options. Without any visible consultation amongst themselves the paint industry negotiators in most countries have lobbied their legislatures to avoid specifying a single compliance route. The Industry's strategy is that reduced emissions are desirable but it wishes to see regulations framed around air quality standards or emission limit value expressed as mass of total emissions. This approach then leaves industry free to select the control option which achieves the legislators desire and through this route the best control option will be developed by competitive pressure.

Unfortunately some legislators continue to drive for emission limit values based only on concentration of VOC in the exhaust gases. This can lead to approval for high mass of emission from processes which require high air flows whilst penalising processes with low mass emissions, which need only very low air flows. The concentration approach also ignores the great benefit that results to the atmospheric emissions when low VOC paints are used. No doubt the debates and lobbying will result in improved legislative focus in due course.

The UNECE VOÇ Protocol lists four different control options to reduce atmospheric emission and the paint industry has selected from all four options depending on the needs of particular industry segments:

(i) Substitution of VOCs.

Great progress has been made in formulating alternative coatings which give very high standards of performance whilst containing low levels of VOC or in some instances no VOC. Research in the industry has generated several alternative approaches to low VOC coatings some of which are particularly useful for the needs of some sectors yet quite inappropriate for others.

Radiation cured coatings contain reactive diluents which reduce the application viscosity but which are reacted into the polymer structure during the curing process thus substantially reducing the emissions to atmosphere resulting from the use of conventional solvents in 'conventional' paint systems. However, radiation cured systems require fixed energy sources and cure rates vary according to the distance between the energy sources and the substrate. Thus, whilst these systems can result in reduced VOC emissions, their use is largely limited to the coating of sheet or flat web substrates in industrial processes, and is certainly not appropriate for coating car bodies nor for decorative air drying coatings.

For most sectors of coatings usage it is possible to formulate systems of higher solids content and this is being done. It appears likely that for most sectors it will not be possible through this route to meet the regulators demands for reduced emissions in full whilst maintaining application properties and film performance characteristics. Nonetheless in some sectors substantial progress will be made and others, for example for the *in situ* coating of bridges in cold, damp climates, the use of high solids systems may be the compliant technology which comes closest to meeting the overall practical needs of society.

The use of powder coatings has been growing at around 15% per annum over the last decade or so, and whilst in many areas of use powder coatings can offer significant cost advantages and/or film performance benefits, it is clear that much of this growth has also been stimulated by concern over solvent emissions. Powder coatings are the ideal compliant coatings for many industrial processes especially the coating of metal and heat resistant plastics and boards and their rapid growth rate reflects this exceptional capability. The need to stove these materials at relatively high temperatures prevents their use in air drying markets, and limits their use on temperature sensitive substrates such as wood and plastic films. The relatively high film builds that are necessary for good film coalescence has limited their use in can coatings and improvements in appearance, durability, and colour styling capabilities still need to be demonstrated before wide acceptance can be expected as car body coatings.

Water-borne coatings are another potentially compliant technology that has grown in market share quite markedly in the last two to three decades and again much, but not all, of this growth has been driven by concern over solvent emissions. Some major areas of use include electropaint technology for automotive primers and industrial uses, water-borne spray technology for coatings for two piece cans, automotive spray surfacers and colour coats, and a wide range of decorative paints. It is, though, still difficult to find a truly full gloss high durability air drying coating and difficult to foresee a means of using water-borne coatings to repaint the exterior of ships dry docking during winter in North Europe or Korea.

Whatever the original drivers it is obvious that without these developments in coatings technologies, the emissions of VOCs to atmosphere would have been much greater.

The remaining three options listed in the UNECE VOC Protocol can also apply to the manufacture and/or use of paint:

(ii) Reduction by best management practices.

Techniques in this area include a lot of items of good sense which can be justified on economic as well as on environmental grounds.

At their most formal they can amount to a fully speciated and quantified mass balance within a solvent management plan, with clear objectives for reducing both fugitive and contained emissions. At their most simple they focus on replacing the lid on a paint tin when it is not in use. It is really quite surprising how solvent emissions can be reduced by always putting lids on portable mixing tanks in paint factories and closing lids in paint thinning tanks at coater's factories. The design of storage tanks so that head space fumes can be returned to road delivery tankers and rigorous preventive maintenance on all flange joints and pumps can also make substantial reductions in VOC losses.

(iii) Recycling and/or recovery for re-use.

Most paint manufacturers and paint users have made progress in reducing the volume of solvent used for cleaning purposes under their application of best management practices and some will collect the remaining usage for redistillation and re-use.

There are also many users of paint in contained processes who collect the solvent evaporated during the drying stage by adsorption onto an activated carbon bed or by absorption in some form of proprietary oil for subsequent recovery and re-use. Due to the nature of the VOCs used in paint manufacture, condensation techniques are not widely used though they do find wide acceptance as techniques in other industries.

(iv) Destruction by post use control techniques.

These techniques can usually only be applied to the manufacture of coatings or their use in contained conditions.

Biological treatment of VOC loaded air streams can be a very effective system for compliance with emission legislation especially when the VOC analysis is of constant composition and where the discharge is both continuous and of relatively low concentration.

However in paint manufacture and in paint usage these ideals are met only rarely and under these circumstances the most commonly used control technique is incineration. This can take many different forms. Catalytic incineration allows combustion to take place at lower temperatures whilst straight thermal combustion involves the lowest capital cost. Regenerative and recuperative incinerators can each show benefits in capital cost/running cost compromises depending on plant sizes and the emissions challenge. It is not unusual for a recovery system such as carbon bed adsorption to be used as a first stage technique and then for the desorbed gases to be incinerated as this often reduces the need for supplemental fuel in the incinerators.

CONCLUSION

With so many different options available to the industry and with different balances and compromises being of relevance in each sector, the VOC issue is quite complex, and is addressed in part by the regulators and in part by industry. The regulators are increasingly focusing on the definition of the environmental quality standards to be achieved whilst industry is developing the best means of achieving those standards.

Chapter 3

Markets for Coatings

A. R. MARRION

A coating material is a composition applied in a layer to all sorts of surfaces to decorate, protect, or in some other way modify them. Here we are mainly concerned with a class of coatings known as paints which is usually but not always used in thicknesses of a few tens of microns.

As we shall see, a great number of different paint compositions are in use in a great many markets. They have in common a skeleton composition of binder, pigment, and solvent or diluent (though the last two are sometimes omitted), and the fact that they are at some point in their lives liquids (if only transiently) and later, solids.

Coatings businesses are usually characterised according to the substrates coated. If an industry produces objects which it wishes to coat, it will be served by a coatings supplier who will probably formulate compositions to fulfil its needs as he sees them. The extent to which his perceptions are correct is measured by his success in securing and retaining the business.

Indeed, paint formulators must acquire an intimate knowledge of their customers' industries in order to solve the many problems which will arise in the development of a promising material, and the many more which are likely to manifest themselves when it is used under real conditions. They will wish to know what the substrate is, whether it is compatible with the proposed coating, how the coating is to be applied, and what the scale and economics of the process are likely to be. The existence of special constraints, for example on toxicity or volatile organic emissions will also be taken into account.

If our manufacturer encounters a coatings supplier whose products make his life easier, or save him money, his gratitude is likely to be tangible, and the basis of a prosperous and lasting relationship. (The more so if the technology is suitably protected by patents.) The impact of developments in antifouling marine paint was discussed earlier. It is an excellent example of a technology-driven business development. The increasing recognition of the possible environmental impact of coatings materials was also discussed earlier and is providing a rich vein of opportunity for innovators; the high energy efficiency and virtual absence of organic volatiles from powder coatings is ensuring rapid growth in their markets. In the same way, water-based coatings are creating great interest and taking parts of the industry by storm.

Some examples of different coatings businesses are discussed below. The list is confined to a selection of conventional paint markets since the intention is to provide a flavour of the commercial constraints acting on paint chemistry. No consideration is given to the extensive commerce which exists in coatings for several types of vehicle, wooden and steel furniture, cloth, paper, leather, electronic circuit boards, roads and airstrips, fibre optics, photographic 'film', the human body, and so on.

It is clear that most objects painted are made from the familiar materials of construction such as steel, wood, aluminium, concrete, and plastics. However, it is probably true to say that the most popular substrate is paint itself, since most successful schemes are multi-layer, often incorporating a primer whose function is to adhere to the workpiece, fill minor imperfections, and provide a base for succeeding coatings.

Top coats by contrast exist at the interface with the environment and must provide resistance to the conditions encountered such as irradiation, oxidation, hydrolysis, and the action of aggressive fluids as well as the required cosmetic features such as colour and gloss.

The most marked distinction between coatings markets is made by the possibility of using stoving processes or otherwise.

Small metal objects (as large as car bodies) which are produced continuously on a production line lend themselves to rapid curing at elevated temperatures. The coatings so produced are superior to those which have hardened at ambient temperature and indeed only such a rapid process is compatible with automated manufacture.

Many objects on the other hand are too large or too sensitive for stoving and room temperature drying coatings are applied. Ships and lips might serve as examples.

Inspection of the following cases may make the distinctions clearer.

Buildings

Buildings of one sort or another account for about half the coatings materials sold worldwide. They include concrete, plaster, brick, wood, steel, galvanised steel, and plastics substrates, some of which are exposed to the weather, and some of which are protected.

Concrete and plaster present porous, alkaline substrates. Consideration has to be given to the movement of water and salts and the tendency of sensitive organic groups to be hydrolysed. Polyester paints, for example, are not used on alkaline substrates in moist locations.

Wood is also porous and inclined to expand and contract with variations in humidity. Brittle paints, or those which embrittle with time such as the traditional alkyds, are likely to have a rather short lifetime on exterior woodwork unless particular care has been taken in the selection and pretreatment of the timber. More flexible alkyds have been developed over recent years and show superior performance, but a more effective treatment may be the application of a biocidal preservative without real film-forming character.

Alkyd based paints have traditionally been used on most parts of buildings, but they are increasingly being replaced by latex paints which dry by loss of water (the 'emulsion' paints familiar for many years on interior walls and ceilings). Such materials are applied by brushing or roller except on large industrial buildings where airless spray allows rapid coverage.

Steel Structures

Large land based steel structures such as bridges, pipelines, tanks, and other chemical plant are most frequently coated with epoxies pigmented with micaceous iron oxide, for example, to give maximum corrosion resistance. They are in some cases overcoated with alkyds or two pack urethanes to provide enhanced weatherability and the required colour. (A tank may be painted white to protect its contents from undue solar heating.)

Again, water-based systems are making significant inroads, subject to difficulties in getting them to form satisfactory films under humid conditions.

Bare steel, before fabrication, is often thinly coated with 'etch' or 'wash' primers, or zinc silicate primers and the emission of harmful fumes during welding operations is a significant consideration.

Steel structures for the most part require of their coatings good resistance to weather and a few bold colours. They are not much subject to mechanical abuse such as friction. When galvanised, as in the case of electricity pylons, they are allowed to weather for many years before painting. Large amounts of coating may be consumed in one project and the emphasis is very much on cost effectiveness. They are usually applied by airless spray.

Ships

Ships and static marine structures, such as oil drilling platforms, have the same general requirements as their terrestrial counterparts, though the potential for vastly increased corrosion by sea water is a significant factor. Some parts are also subject to severe abrasion due to anchor cables, contact with jetties, contact with ice in polar waters, and the effect of airborne ice particles. The effect of abrasive cargoes, such as coal, iron ore, or pig iron, is largely confined to the insides of ballast and cargo tanks which also have to contend with an unpredictable succession of aggressive fluids ranging from fuel oil and sea water to chemicals and foodstuffs.

Tank interiors are coated with densely crosslinked epoxies. Other surfaces are coated with materials often superficially similar to those used on land-based structures, but formulated to withstand the more severe conditions they encounter. They are applied by airless spraying, often at considerable thickness.

Yachts

Yacht coating offers a striking contrast to that of large ships since yachts are objects of luxury, often painstakingly coated by their owners. Economy is given a lower priority than appearance and weather resistance. Timber surfaces are often varnished to retain and enhance the beauty of their grain. Glass-fibre composite hulls are provided with a coloured 'gel coat' in manufacture, but are often painted to change their colour or improve their gloss.

Alkyds of high quality and two pack polyurethanes are most often used and are applied by brush or conventional (air assisted) spray.

Aircraft

Aircraft exterior coatings are amongst the most severely tested in service. Flight at high altitude subjects them to intense ultraviolet irradiation, salty air is liable to initiate corrosion which may be promoted by mechanical stress, rapid temperature changes may cause cracking, and (as if that wasn't enough) leaks of phosphate ester hydraulic fluid have to be reckoned with.

It is usual practice to employ an epoxy primer, heavily pigmented with a leachable chromate to passivate any metal exposed by minor damage. The top coat is an isocyanate cured polyester for maximum weatherability and chemical resistance. The recent introduction of fluoropolymers to such compositions has provided further enhancements in performance.

Paradoxically, it is necessary to strip the coatings from aircraft at intervals to inspect the airframe. It must be done very rapidly to minimise time out of service, and with agents which do not attack the metal. Efforts have been made to develop coatings which possess all the above attributes whilst being easily strippable.

Motor Cars

Car plants are nearly always very large scale operations—the archetypal production lines. Bodies are manufactured continuously and undergo coating processes before being built up into the finished vehicles. Car owners demand a very high standard of appearance and resistance to corrosion, and most manufacturers use cathodic electropaint as a primer which can provide even coating inside recessed areas such as doors. It is covered by a relatively thick surfacer coat to hide imperfections in the metal and provide a smooth substrate. The top coat may be a melamine cured polyester or acrylic, or less frequently a thermoplastic acrylic, colour matched to the current fashion with the greatest care.

Metallic finishes found their first use in the Automotive Industry where they contributed materially to body design and they have given rise to the widespread adoption of 'clear over base' systems where a thin pigmented thermoplastic layer is overcoated with a clear, often in a 'wet on wet process'.

Automotive top coats are electrostatically sprayed, often by means of a 'high speed bell', a spinning disc onto which the paint is fed to be charged and atomised. They are stoved at about 130°. The earlier coats are stoved at somewhat higher temperatures so that they are not much affected by later heat treatment.

Car Refinish

Car refinishers operate much smaller concerns than original equipment coaters. In some cases they are one-man operations using hand spray guns, and buying just enough material to coat one car at a time. The suppliers ability to match any of the enormous number of colours on the road is of overriding importance, and it is necessary to be able to obtain high gloss under conditions which may be less than ideal. It is impossible to stove the car, which is now equipped with all sorts of heat sensitive items, from rubber tyres to the sun-glasses in the glove-box. However it is often possible to provide a 'low-bake' or 'forced cure' below about 100 °C.

Traditional refinish paints were based on nitro-cellulose and dried by evaporation of solvent. Being rather soft they lent themselves to polishing to remove defects. More recently however, two pack systems based on hydroxyacrylics cured with isocyanates have gained popularity since they can achieve a high gloss 'straight from the gun'.

Food and Beverage Cans

Tinplate cans for food or drink need to be lined with a continuous film of highly resistant material if they are not to be attacked by their contents with resultant metallic taints or even leaks. It is also necessary for the coatings to withstand considerable deformation without failure, to be free of any extractable components which might themselves taint the contents, and to be recognised as acceptable for food contact. Similar constraints apply to inks used on can exteriors.

Various epoxy compositions are used in most cases, and stoved at temperatures in the 250–300 °C range for a few seconds, consistent with the very high production rates of modern can lines. As in other markets, water-based can lacquers are rapidly gaining in popularity, and are typically based on high molecular weight epoxies modified with acrylics.

Traditional three piece cans were made from tinned steel strip, roller coated in the flat state then formed into a cylinder with soldered seams before tops and bottoms were fitted.

More recently, the need for economy of materials (and the high performance of coatings) has led to the development of two piece cans with bodies extruded from single tinplate coupons. The walls of such cans are much thinner than in three piece cans, and have much thinner, probably discontinuous, tin layers so that still greater reliance is placed on the organic lining. The need to coat the inside of a metal cup uniformly and with an extremely low incidence of 'holidays' (holes), in a fraction of a second places enormous demands on application technology. It is normally achieved by an automated spray arrangement though other systems are under active consideration.

Coil

Coil coating is the application of polymer film to steel (or aluminium) strip prior to fabrication. Application is by roller, and the strip is fed at high speeds through high temperature ovens. The process lends itself to automation and is usually carried out by the manufacturers of the metal stock.

The coating must come unscathed through the deformation involved in subsequent fabrication, and arrangements must be made to minimise the effects of corrosion at cut edges. Coated coil is ideally suited to the manufacture of rather flat objects such as domestic appliance panels, garage doors, and cladding for buildings. For exterior use, extreme ductility has to be reconciled with high weather resistance.

Poly (vinyl chloride) plastisols enjoy wide popularity, but for highest endurance under sunny conditions, fluoropolymers are unrivalled and are gaining ground.

There exist, as mentioned in the preamble, many other markets to consume large or small amounts of paint, and many new ones will appear in the future. To a greater or lesser extent all are dependent on the coating for the quality of their product and the success of their industry.

Chapter 4

The Rheology of Coatings

P. A. REYNOLDS

INTRODUCTION

If a film is to function effectively as a barrier it must be continuous and free from defects. Of necessity, such films formed from paints require that the paint exists in the liquid state at the point of application.

One of the principle reasons is that the discrete regions of a coating, such as spray droplets for example, must coalesce to form a continuous film. Moreover, the material needs to be handleable; it must be easily transferred from its container to the surface and applied at the required film thickness. Although this process does not require the material to be liquid, conventional paints are so under normal conditions. The post application behaviour of a coating, on going from a liquid to a solid must be such that it 'holds up' on the surface without running or dripping and that the visible surface imperfections and undulations 'flow out' to give a smooth film of the required thickness. The principle property of the paint common to all these processes, and of critical importance, is the way in which the material flows. The science of the way in which materials flow and deform is called **rheology**. We can see that paint rheology is not a simple flow property when we recall that after application the paint must hold-up, possibly on a vertical surface, yet 'flow out'.

For these two effects to occur would seem unreasonable since they appear to be mutually exclusive. However, most coatings are formulated such that the rheology adequately accommodates both requirements. In this chapter the important issues are discussed and illustrated.

Before proceeding to discuss the specific nature of the rheology of paints and its influence on the preparation and application of coatings, let us consider rheology, the science of deformation and flow of materials, in more general terms.

RHEOLOGY

'Viscosity' is the word most often used to convey an impression of the ease or degree of difficulty encountered when attempting to cause a material to flow. In a kitchen, for instance, a chef would probably use the word 'consistency' to convey exactly the same meaning. The word consistency has an immediate impact on both novice and expert, and in the context of its general use, rejoices in being sufficiently vague to ensure its usefulness. A contemporary definition of consistency for rheologists is 'a general term for the property of a material by which its resists permanent change of shape' whilst older definitions add '. . ., defined by the complete flow-force relation'. A dictionary definition talks of the '. . . degree of density, firmness, or solidity, especially of thick liquids'. Thus we tend to use adjectives such as thick and thin to describe fluid properties. For example we often talk of fluids being water thin.

Conversely viscosity is a precisely defined quantity which relies on rigidly defined, measurable parameters. It still, however, is associated with the ability of a material to resist a deformation or flow. In order to appreciate the rheology of paints in general and viscosity in particular it is necessary to define the measurable parameters systematically. Table 4.1 gives definitions of some common rheological terms which will be used in the following text.

Shear Flow

The easiest type of flow to define and control is (simple) shear flow.

This is demonstrated in Figure 4.1a where we can see that a force applied to the top surface area of a cube of material produces a deformation in that surface and gradually diminishing deformations in what can be envisaged as layers in the remainder of the material. This force, F, per unit area, A, is termed the shear stress (σ) and has been shown to create a deformation characterised by the angle γ known as shear. If this shear stress were maintained with time on the surface of a fluid then the angle γ would change with time, however the rate of change of γ would be constant, and this is known as the shear rate, denoted $\dot{\gamma}$ (the dot denotes a time derivative). Thus, essentially the shear rate is a measure of how rapidly a material is deformed, or made to flow.

The viscosity (η) of a material is defined as the ratio of the applied force per unit area to the shear rate

$$\eta = \frac{\sigma}{\dot{\gamma}} \tag{1}$$

Table 4.1 *Some common rheological definitions*

Anti-thixotropic	A slow fall, on standing of a sample, of a consistency gained by shearing.
Apparent viscosity	The shear stress divided by shear rate. It is not a constant coefficient.
Coefficient of viscosity	The shear stress divided by the shear rate. A constant, defining a Newtonian fluid.
Consistency	A general term for the property of a material by which it resists permanent change of shape.
Dilatency	An increase in volume caused by shear. (Often confused with shear thickening.)
Newtonian fluid	A fluid for which the shear stress is proportional to the shear rate.
Non-Newtonian fluid	A fluid for which the proportionality between shear stress and shear rate is not constant with shear rate.
Plastic	A material which flows when a yield stress is exceeded.
Pseudo-plastic	Shear thinning. Often used in a context where shear stress is linear with shear rates at high shear rates, but no yield stress can be detected.
Shear	The change of angle in a deformed body.
Shear rate	The change of shear per unit time.
Shear thickening	The increase in viscosity with increasing shear rate.
Shear thinning	The decrease in viscosity with increasing shear rate.
Shear stress	The force per unit area parallel to the area.
Thixotropy	A slow recovery, on standing of a sample, of a consistency lost by shearing.
Viscosity	The resistance of a material to deformation. The shear stress divided by shear rate.
Yield stress	The stress which must be exceeded for a non-recoverable (viscous) deformation to result.

The unit of shear stress is Newtons per square metre (Nm^{-2}) and that of shear rate is per second (s^{-1}) (rate of change of an angle with time). The unit Nm^{-2} can be written as Pascal (Pa) which divided by s^{-1} gives the unit of viscosity as Pa s. Table 4.2 gives examples of the viscosity of some simple liquids.

Newtonian Fluids

Given these three defined properties for simple shear flow a large set of material behaviours can be described.

The simplest rheological behaviour a fluid will show occurs when an applied stress is proportional to the resulting shear rate or *vice versa*. The constant of proportionality between the two, is the coefficient of viscosity, η, as in Table 4.2. Because viscosity is invariant with either variable in this

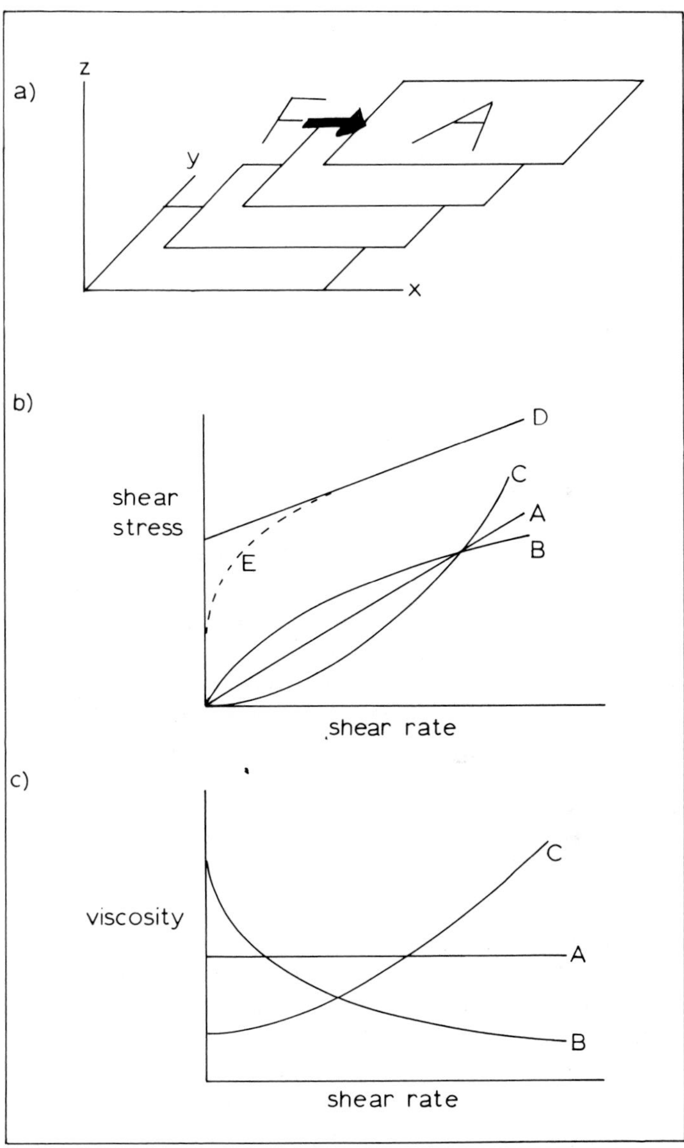

Figure 4.1 (a) *Schematic diagram of simple shear flow;* (b) *Shear stress–shear rate behaviours*—A. *Newtonian,* B. *Shear-thinning,* C. *Shear thickening,* D. *Plastic,* E. *Pseudo-plastic;* (c) *Viscosity–shear rate behaviour*—A. *Newtonian,* B. *Shear thinning,* C. *Shear thickening*

Table 4.2 *Examples of the viscosities of some materials*

Material	*Viscosity*/Pa s	*Temperature*/°C
Acetone	0.3×10^{-3}	20
Water	1.0×10^{-3}	20
Ethanol	1.2×10^{-3}	20
Propanol	2.3×10^{-3}	20
iso-Propanol	2.9×10^{-3}	20
Diethylether	1.7×10^{-3}	−100
	2.8×10^{-4}	0
	1.2×10^{-4}	+100
Air	1.5×10^{-5}	20
Liquid air	1.7×10^{-4}	−192
Glycerol	1.5	20
Glucose	9.1×10^{12}	22
	2.8×10^{8}	40
	2.5×10^{1}	100

case we say that it obeys Newton's law (postulate) and the material is Newtonian. This is shown in Figure 4.1b as a line of constant slope of shear stress with shear rate passing through the origin, and in Figure 4.1c as a line of zero slope and constant viscosity with shear rate. Many simple, low molecular weight materials, are Newtonian when liquid. Note here that viscosity changes with temperature, thus materials generally exhibit lower resistance to deformation when hot than when cold.

Non-Newtonian Fluids

For polymers, polymers in solution, particles in a continuous phase, or indeed both polymers and particles in a solvent (continuous phase), it is common to find that the viscosity is no longer invariant with either shear rate or shear stress. These types of material are termed non-Newtonian. Figure 4.1b and 4.1c attempt to illustrate the general types of behaviour often observed.

The most common behaviour is shear thinning where with an increasing shear rate the material has a decreasing viscosity. Paints and coatings will generally show this type of behaviour and it will be noted that they usually contain both polymers (in solution) and pigment particles. Clearly, their absolute behaviour in terms of how much thinner they become and in what shear rate region this occurs, will depend upon the exact chemical and physical nature of the materials used.

The converse of shear thinning is shear thickening, whereby the viscosity is increased by increasing the shear rate. This behaviour can be observed with materials containing a high concentration of particulates, pigments for example. The coating industry often exploits this behaviour in paint manufacture by using 'mill-bases'. Mill-bases are highly pigmented mixtures, exhibiting shear thickening, and hence when caused to flow rapidly show very high viscosity.

Aggregates of particles are in close proximity to each other hence the difficulty in moving relative to each other when in high shear flow gives rise to high viscosity. This action also leads to mutual attrition which disperses them into smaller particles.

Another class of non-Newtonian behaviour is that of plastic flow. This type of flow requires that a finite stress is applied to the material before flow is possible. That is, the material exhibits a yield stress, as demonstrated in Figure 4.1b where the curve for plastic flow intersects the shear stress axis. Very often materials, when studied at higher shear rates, appear to show a form of shear stress–shear rate behaviour which would appear to extrapolate to zero shear rate and result in a yield stress. However, when measurements are made at successively lower shear rates the curve tends towards the origin. This is called pseudo-plastic flow for obvious reasons, but is often used synonymously with shear thinning. In reality it is a class of shear behaviour.

Time Dependency

The remaining variable which can have a profound influence upon the rheology of a material in simple shear flow is time. It is often found that at a constant shear rate the viscosity of a material continually changes for a period of time prior to settling to a steady-state value. If, after the sample has been left to 'recover' for a significant period of time, the behaviour is seen to be repeatable, that is the original steady-state value is not immediately measured, the material is showing time-dependent behaviour.

Should the viscosity decrease with time, and thereafter increase to return to its original value after a period of 'rest', the behaviour is termed thixotropic. Often confusion arises in the use of the term thixotropic since it is equated with the decrease in viscosity whereas it is the rebuilding of the 'structure' of the fluid to recover the initial high viscosity which is the thixotropic nature.

Conversely, an increase in the viscosity during shear which is lost during a period of rest is termed anti-thixotropy. This is also commonly termed rheopexy today, although the original usage of rheopexy was not synonymous with anti-thixotropy.

Thixotropy is far more commonly found than anti-thixotropy. Very often this time-dependent behaviour is tested for by observing the viscosity variation with increasing and then decreasing shear rates. This results in a thixotropic loop as shown in Figure 4.2a. Given the preceding test, true thixotropy could only be confirmed if, after a period of rest, this loop could be repeated.

In reality this thixotropic loop is a two dimensional view of a three dimensional surface composed of the time-dependent viscosity at constant shear rate, Figure 4.2b, and the shear rate dependent viscosity taken at a constant time after the commencement of shearing (Figure 4.2b). The thixotropic loop is seen as the trace of behaviour over the resulting surface (Figure 4.2c).

This rebuilt viscosity for thixotropy, or decayed viscosity for anti-thixotropy is often spoken of as the 'structure' of the material. The idea to be conveyed here is that the flow degrades (or builds) some micro structural units within the material which can be regained (or lost) on standing.

RHEOLOGICAL MEASUREMENTS

For determination of the absolute properties, within well defined flow processes, simple shear flow for instance, a large number of commercial instruments is available. They are generally rotational devices in which the geometry of the device holding the sample is defined so that one part of the geometry is in rotation, whilst the other part is static, and this creates a well defined and constant shear rate throughout the sample.

It may be that the instrument sets the shear stress (actually torque) and measures the rotation rate or *vice versa*. Irrespective of which of the two quantities is set and which is measured, shear rate and shear stress are known, and thus a material's viscosity profile with shear properties can be determined. Curves similar to those shown in Figure 4.1b and 4.1c result.

Rheology and Coatings Reality

In rheology, we work hard to simplify the interpretation of information and in order to do this we use more and more elaborate instrumental methods. These produce well defined flow regimes in which simple shear flow is generated. However simple shear flow is not generally established in any real application process nor can it be considered to be established in other flow phenomena associated with coatings.

The processes designed for the application of coatings are based on other criteria than simple flow. Thus the common methods of application, brush, spray, roller, and electrodeposition, for instance were developed

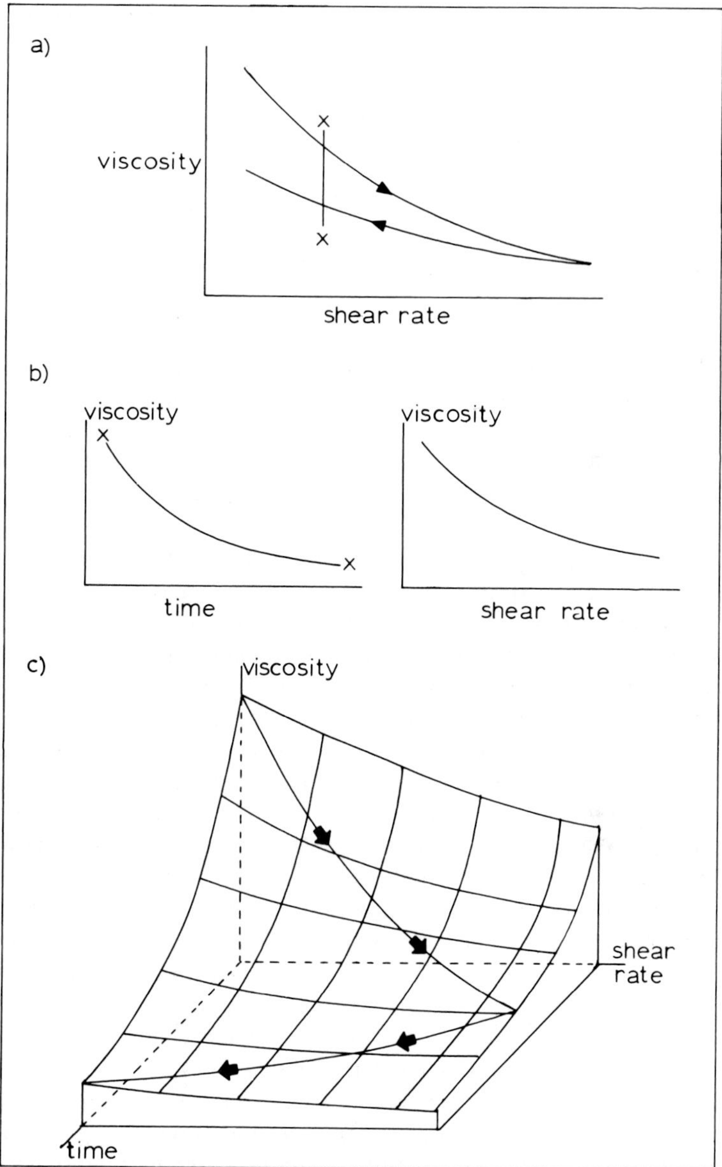

Figure 4.2 (a) *Typical thixotropic loop. Arrows indicate the direction of the shear rate sweep;* (b) *Viscosity–time at a constant shear rate and viscosity–shear rate behaviour of a thixotropic material;* (c) *Trace of the behaviour of a thixotropic material on a viscosity–shear rate–time surface*

with consideration of criteria such as speed of application, mechanical simplicity, transfer efficiency, and ease of equipment cleaning amongst a host of others. Given this mismatch we attempt to relate parts of a process to the extent of shear generated in a process.

Clearly the use of simple shear viscosity is somewhat flawed since we have to work hard at obtaining such a flow. The reality is that we can ultimately only give an estimate as to what the equivalent shear rates may be in a real process.

Industrial Methodologies

Within the paint industry many methods and techniques have been developed to relate a 'flow quantity' to a particular coating performance characteristic, so for a particular application an industry standard will be developed. It is beyond the scope of this text to review such methods but a quick consideration of some of these techniques reveals some interesting points.

One such industry standard is the Ford No. 4 efflux cup. This is essentially a machined cup of defined dimensions having a conical bottom to the cup at the centre of which is a hole of a defined size. To take a reading the orifice is sealed with a finger and the cup is filled with material with the excess caught in an overflow trough. The finger is removed and the time for the material to stream out measured. This is a measure of the 'viscosity' having the units of seconds. It is effectively a comparative way of describing the ease with which a material will flow in a given environment.

An experienced formulator is able to assess from this information how a coating will perform, for example, in a spray application.

It will be readily appreciated that this type of measure is suitable for fluid (thin) materials, which are essentially Newtonian, thus a definition of shear rate is unnecessary. It ought to be equally apparent that as a measure for highly non-Newtonian materials the method has little use since it can give no information regarding the degree of shear thinning and how this will effect the actual application.

Another industry accepted standard instrument is the Stormer viscometer which describes the viscosity of a material by measuring the mass required to maintain a paddle immersed in the material in rotation at a constant 200 r.p.m. The falling weight drives the paddle through a series of gears and pulleys and the rate of 200 r.p.m. can be quickly established with a stroboscope. The unit of viscosity is a derivative of the variables and is expressed as the Krebs Unit (KU), which is a unit specific to this instrument. The effect of measuring materials which show a time-dependence is even more serious than with the Ford Cup. Thixotropic paints for instance, will show a continuing acceleration of the paddle as the paint is reduced in viscosity by its action.

With imagination such measurements can be seen to relate to the rheology of a paint and be relevant to mixing, thus manufacturing perhaps.

As a formulation aid the 'Rotothinner' viscometer is a useful tool since it mixes the paint at the same time as making viscosity measurements. It has generally been acknowledged that for paints to be applied by spray, brush, or roller they require a viscosity in the range 0.05–0.5 Pa s. Thus this instrument allows formulation changes to be made whilst measuring the viscosity. The principle of operation is simple in that a disc shaped stirrer is immersed in a standardised container of paint and rotated at constant speed. The torque transmitted though the sample is measured via a spring connected to the paint container.

Those devices mentioned above are not capable of defining a shear rate clearly. The ICI Cone and Plate viscometer was designed with an acknowledgement that coatings are generally shear thinning and that brushing, spraying, or roller coating of paints takes place at high shear rates, around $10\,000\,s^{-1}$. The device uses a small diameter wide angle cone rotated at a constant speed, in contact with a plate. This accurately defines the shear rate throughout the sample and is set to give a shear rate of $10\,000\,s^{-1}$. The viscosity is again measured through a calibrated spring torque device.

With unknown materials these estimates of paint rheology are of little value. However, under the conditions of use for which they were designed they are very often ideal tests for determining the optimum rheology for a material (of known generic form) in a particular process.

RHEOLOGICAL PROCESSES ASSOCIATED WITH COATINGS

Once in possession of a knowledge of the rheological behaviour of a material in shear flow, the task then becomes one of determining which parts of the curve, shear rate-viscosity or thixotropy for instance, can be identified with various processes or phenomena. Thus brushing, roller coating, and spraying as well as shears generated in manufacture and storage can be identified with particular regions of the overall characterised behaviour.

In order to appreciate the flow or rheological characteristics associated with a paint consider how a paint proceeds from manufacture to a final dry film. Generally, a paint contains four generic components: polymers or resins, pigments, solvents, and a variety of additives. For powder coatings there is no associated solvent but the thermal properties of the polymer ensure a degree of fluidity at the elevated temperatures used in curing the coating. Each component has a dramatic effect on the rheology of the material at each stage of its life.

Manufacture

At the paint manufacturing stage the principle action, apart from producing a homogeneous mixture, is to reduce the size of the agglomerated pigments. This involves mechanically breaking the primary particles apart, wetting their surfaces, and creating a different surface such that they do not reaggregate. Most conventional formulations rely on the binder (resin) to wet the pigment particles. As a result, considerable mechanical energy is required to 'mill' agglomerated dispersion to a 'smooth' paste, as previously mentioned often a shear thickening mill-base.

Many types of mill are commercially available to supply the energy required via different mechanical routes. They range from the very simple, such as the high speed disperser which is essentially a high speed rotating disc with sawteeth at its periphery, to the more complex such as colloid mills or bead mills. The high speed disperser although simple sets up flow patterns relying on a shear thickening mill-base for its greatest efficiency. Thus nearest the blade the high shear disperses the agglomerates as well as inducing the self attrition mechanism, whilst the lower viscosity away from the blade ensures the mobility of the material necessary to reintroduce mill-base into the dispersing region continuously.

Bead mills, or sand mills, have chambers filled with small beads or sand and made to move rapidly by a series of rotating discs. The chamber is small compared to the volume of mill-base which is made to flow through the chamber. Here the shearing action of the beads in motion 'grinds' the pigment aggregates. A screen allows passage of the milled paint but retains the beads. Multiple passes through the mill can be made.

It turns out that rheological characteristics of a mill-base which is of optimum efficiency in one type of mill may be such that it may not be milled effectively in a different type. Mill-bases must be specially formulated for individual types of mill.

Most often the paste-like dispersions are combined by a low energy mixing method with the remainder of the formulation, and a paint of the correct 'consistency' results.

The finished paint is discharged into cans and stored.

In-Can Stability

Since the storage time can be quite considerable the 'In-Can' stability of the paint is important. The principle problem here is that the discrete, but small, pigment particles have densities far in excess of the continuous medium in which they are immersed, and hence have a high propensity to settle to the base of the can.

Such sediments can be extremely cohesive and difficult to redisperse even though the individual particles are stable to aggregation. It is easy to envisage that this settling is a low shear and slow rate process but none the less, because of the time scale involved, a problem. In order to prevent settlement the paint is often structured with additives, so that a high viscosity prevails at low shears. Structured paints often exhibit thixotropy or a yield stress value.

Application Process

Because paints are most frequently either highly shear thinning or thixotropic, they tend to aid the application process. High shear rates generated in these application methods destroy the inherent structure within a paint, for the duration of the application, which considerably increases the ease with which they flow in such processes. For example, brushing is considered to generate a shear rate of approximately 10^4–$10^6\,s^{-1}$, that is, high. Hence if we wanted to relate the brushing characteristics of a series of paints to their rheology, we would look at and compare their high shear viscosity.

An ICI cone and plate viscometer would be best suited for this measurement. The same argument would apply to spray coating methods, both air assisted and airless. The main problem with adjusting formulations based on high shear viscosity measurements is that this is an insensitive region of the viscosity–shear rate curve. Small variations in this region tend to be magnified many fold in the low shear region.

Another major consideration that has to be taken into account when adapting the rheology for an application process is that in order to apply a sufficiently thick coating to the substrate the high shear rate viscosities also need to be relatively high. This is a necessary condition since otherwise the stresses (gravity) acting on a thick dense paint film would cause it to run or sag.

This condition subsequently imposes a limitation on the low shear viscosities, since affecting change in the high shear region also produces change in the low shear region.

Similar limitations are imposed on industrial roller coating applications. Here large rollers, and they may be up to several metres wide, are used to lay a coating onto metal sheets or coils, for instance. The principle is that a pick up roller rotates in a bath of the coating which ultimately, via several other rollers, transfers the coating at a well defined thickness to the metal. Thus the rheology, amongst other considerations, determines the film thickness, whether the coatings will fly off the rapidly rotating rollers, and the extent to which surface imperfections in the coatings will flow out prior to the thermally induced curing. Often a Ford cup measurement is used to determine the 'viscosity' of the material for application. If the final application roller is rotating in the opposite sense to the direction of travel of the metal sheet or coil it is termed reverse roller coating. When rotating in the same sense it is termed forward roller coating. Reverse roller coating is a high shear process and minimises surface defects in the coating associated with it. Forward roller coating is a relatively lower shear rate process which sets up an inherent instability seen as ribs of coating in the direction of travel. These must flow out and level prior to curing.

Table 4.3 identifies some typical processes and estimates the equivalent simple shear rates associated with them.

Table 4.3 *Viscosity–Shear rate influence on paint properties*

Approximate range of shear rates/s^{-1}	*Application*	*Required viscosity*
10^{-6}–10^{-2}	Prevention of pigment and particulate settlement	High
10^{-2}–10^{-3}	Promotion of flow-out and levelling	Low
10^{-2}–10^{-3}	Sagging and slumping control	High
10^{0}–10^{2}	Paint pick up on brush	High
10^{0}–10^{3}	Mixing, pumping, and stirring	Low
10^{1}–10^{3}	Forward roller coater application	Low (but depends on required film thickness)
10^{3}–10^{6}	Brushing, reverse roller coating spraying	Low
10^{3}–10^{6}	Control of drag on brush	Low
10^{3}–10^{6}	Control of thick film build	High
10^{4}–10^{6}	Control of paint mist (fly) off rollers	High

Post Application Rheology

Several important changes often occur during and after the application of a paint. If it is atomised it will loose solvent between the spray tip of the gun and the substrate. This results in a lowering of its temperature and an increase in solids content, which will continue during drying. Both of these effects will act so as to raise the viscosity of the paint throughout the entire shear rate region. Once the paint is at 'rest' on the substrate surface its shear thinning nature will ensure a high viscosity in the low shear rate region. Moreover, if the paint is thixotropic it will start to restructure and increase the viscosity.

The importance of these effects is shown below. Prior to significant flow disabling viscosities being reached the paint droplets need to flow to coalesce and form a continuous film. Furthermore, the film needs to flow and level in order to produce a smooth, defect free, glossy finish. These are low shear processes. The time frame in which they occur is important since if the viscosity remains low for too long the propensity of the paint to flow under gravity will be realised and runs or sagging and slumping of the paint will occur. The implication here of course is that a thixotropic material, where the time for restructuring is controllable, will allow significant coalescence, flow out, and levelling to occur prior to runs and sagging becoming appreciable.

Whilst coatings which are roller applied or brushed may form continuous films during application, they still require sufficient mobility to flow-out and level, again with control of the runs or sagging behaviour. Paint so applied is therefore also subject to the influences of shear thinning and thixotropy. Consideration must also be given to film thickness.

Many coatings, particularly those applied by roller, proceed directly from application into a high temperature cure oven, where a chemical cross-linking mechanism is initiated. Several effects are coincident. Because of the increased temperature the viscosity of the paint decreases, probably significantly. Solvent, which may of course be water, will be driven off, leaving the non-volatile polymer and pigments. The viscosity will start to rise accordingly. The cross-linking mechanism will rapidly increase the molecular weight of the polymer which in turn will increase the viscosity very rapidly (with a third or fourth power dependence on molecular weight!). During this thermal treatment, which may be very rapid, sufficient flow must be permitted to produce the required appearance of the films.

Film Defects

There are other effects which can ultimately lead to poor surface appearance and although not caused by the rheology of a coating are facilitated by it.

Flooding and floating are related in that they occur in coatings containing two or more pigments. For flooding to occur one of the pigments preferentially migrates to the surface of the coating giving it a different colour from the bulk of the paint. Floating is similar in that there is a differential concentration of pigments across the surface giving a generally mottled film appearance. This is due to Benard cells. These are small hexagonal areas which are created by convection currents set up by surface tension differences developed during solvent evaporation. They circulate one of the pigments preferentially thus producing a pigment concentration difference across the surface of the paint. Clearly both processes are only possible if flow can occur in the drying coating.

Conversely orange peel, an undulation across the surface of the coating resembling the surface of an orange, is a particular defect which might be eliminated if there was adequate total film mobility during the drying or curing stage.

If the surface of a newly applied film is considered to be a series of peaks and troughs of all amplitudes and wavelengths (within reason) a good analysis of flow out and levelling is the Orchard equation. This describes the decay in amplitude with time $a_{(t)}$, from an original amplitude a_0, of a sinusoidal wavefore of wavelength λ, with a film of average thickness h, and viscosity η.

$$\ln\frac{a_0}{a_{(t)}} = \frac{16\pi^4}{3}\frac{h^3\Gamma}{\lambda^4}\int_0^t \frac{1}{\eta}\,\mathrm{d}t \qquad (2)$$

This tells us that for small wavelengths surface tension, Γ, the driving force to levelling, is sufficient to collapse their amplitudes effectively to zero. However, for large wavelengths or undulations the driving force is not adequate within the time frame of the drying or curing process in which the viscosity is continually rising. That is there is not enough 'flow available'. So the final appearance is typically a smooth film with large wavelength undulations, that are of the order of millimetres.

LOW VOC COATINGS—FLOW PROBLEMS AND SOLUTIONS

In order for coatings to become more environmentally acceptable the amount of volatile organic compounds (VOC) contained within them must be restricted as we have seen in Chapter 2. Removal or partial removal of the organic solvents has a profound impact on the rheology of such coatings. The total removal of solvents from a conventional paint to give a zero VOC paint may be taken as an example. The product will inevitably be highly viscous and possibly enormously shear thinning, which limits its ability to be mixed and pumped during manufacture, limits the formulator's ability to produce a sprayable coating, and limits his ability to produce acceptable surface finishes. One method of alleviating this problem is to use lower molecular weight, hence lower viscosity polymers. This solution tends to be expensive and if taken too far, can yield a highly mobile film before sufficient curing has taken place to retard sagging and slumping. Another way of achieving satisfactory spray and post-application rheology is to heat the paint until the viscosity falls into the correct region. However, a very much increased rate of curing occurs and may take place in the mixed paint prior to use, even to an extent that may prevent its application. Chemistries designed to counter this effect and retard the reaction tend to suffer from lower extents of cure or perhaps no cure at all when used in low ambient temperatures.

Amongst other low VOC technologies, solvent free or 'powder' coatings and water-based systems are most significant.

Powder coatings are paints made from solid polymers, that is polymers whose glass transition temperatures are considerably higher than ambient temperatures and whose fluidity (the inverse of viscosity) only becomes appreciable at temperatures greater than about 100 °C. When fully formulated with pigments and additives these coatings are milled into small particles of about 30 μm average size. The powders are made to flow through a gun which applies an electrostatic charge to the individual particles on their flight to the earthed substrate. The particles adhere not only to the front of the earthed article but because of the change, wrap around the rear of the substrate following the potential field. On heating the coated article in a curing oven the polymer becomes fluid, the discrete particles coalesce, and ultimately flow-out and level. During this process the polymer crosslinks and hence the viscosity rises considerably. This process is shown graphically in Figure 4.3. The most significant feature of cured powder coatings in terms of their appearance is the existence of orange peel. Effectively the coatings do not remain fluid enough for long enough to allow the flow-out of the longer wavelengths or in reality the larger size surface inhomegeneities. These effects can, however, be minimised by judicious formulations.

Water-borne coatings of the latex type present their own special problems which largely result from the very high molecular weight polymer existing in discrete, small particles. These particles are typically

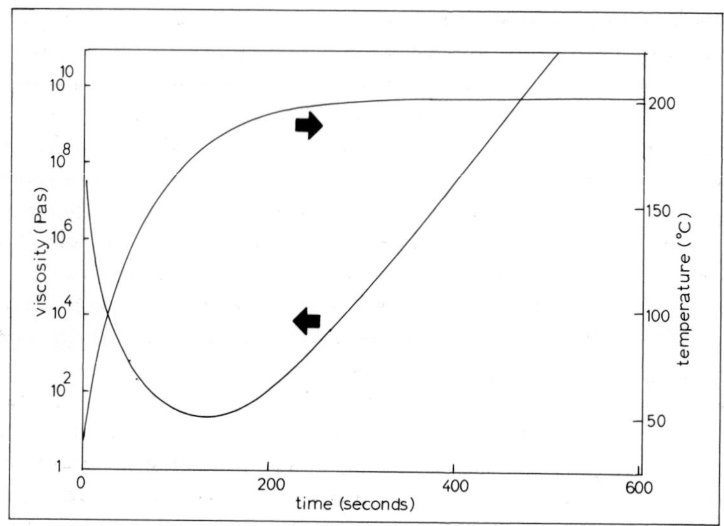

Figure 4.3 *The viscosity profile of a curing powder coating when subjected to a typical cure temperature profile*

of the order of 100 nm in diameter and always less than one micron. In this form they can exist as dispersions of up to 50% volume fraction of polymer and still be 'water thin'. It is often observed that when the volume concentration of latex particles becomes large the viscosity increases very rapidly with small increments in concentration. They display marked shear thinning and can additionally exhibit yield stresses. When formulated with pigments and using realistic concentrations of polymers it is clear that the rheology of these paints has to be modified with various additives.

The rheology modifiers fall broadly into three classes of materials: clay-based thickeners, water soluble polymers of high molecular weight, and associative thickeners. Clay materials work by producing a structure throughout the body of the paint. This structure is sensitive to shear and as a result shows highly shear thinning behaviour and a yield stress. This will give good resistance to in-can settlement. The high molecular weight water soluble polymers form an entangled network at low concentrations effectively thickening the continuous phase (water). By virtue of their high molecular weight they are again highly shear thinning but with no yield stress value. More often they exhibit a zero shear viscosity which, although high at concentrations generally used, does not restrict motion in gravitationally imposed stresses for example. Degrees of flow and levelling can be influenced whilst film build is also considered. Both clays and water soluble polymers can impart degrees of thixotropy to a paint. Associative thickeners are low molecular weight water soluble polymers which have been modified by adding low molecular weight hydrophobic groups such as alkanes. The hydrophobes interact with each other, the hydrophobic latex surface, and other hydrophobic species such as the coalescing solvents and surfactants for instance, to build up a structure throughout the paint. In this way a shear sensitive matrix whose rheological response can be adjusted by changing the chemistry of the associative thickener, the size and number of hydrophobes, the hydrophobicity of the polymer latex surface, and the chemistry and concentration of other hydrophobic materials is obtained. The associative thickeners do not operate by thickening the continuous phase but build individually weak mechanically sensitive contacts which are easily broken and rapidly reformed. As a result the materials do not exhibit thixotropy. Judicious use of these thickeners can enhance or suppress the high shear viscosity for film build and application, the low shear viscosity for flow, levelling, and sag control, and the medium shear viscosity for in-can stability and brush pick-up.

It is not uncommon to use mixtures of different associative thickeners or indeed mixtures of the three different classes of thickeners to produce the desired effect.

CONCLUDING REMARKS

It has been demonstrated in this chapter that the physical process of causing paints to flow, or indeed attempting to prevent flow, for a variety of reasons under certain conditions is not a simple problem. The deformation and flow properties (rheology) of coatings exhibit many and varied forms of behaviour under well controlled, laminar flow conditions. Adapting these properties to give good performance under the constraints imposed in all aspects of the coatings life, from manufacture though storage and application to post application, is one of the naturally difficult problems for a paint formulator to consider. It is often necessary to make a series of compromises to obtain the optimum set of properties. A knowledge of the rheology of paints and how to manipulate it via formulation is a vital tool in the optimisation process.

BIBLIOGRAPHY

The subject of the rheology of materials is a very large area of study in its own right. It is a subject that has its roots in physics, chemistry, applied mathematics, engineering, colloid science, and polymer science. Excellent texts and treatises are available in all these branches but are not for those new to the field. The most appropriate text for the general background to rheology in its most diverse form is that of H. A. Barnes, J. F. Hutton, and K. Walters, 'An Introduction to Rheology', Elsevier, Amsterdam, 1989. This is a very readable book and is highly recommended.

A text more specific to the paint industry is T. C. Patten, 'Paint Flow and Pigment Dispersion', Wiley Interscience, Chichester, 2nd edn., 1979. This takes on the form of an undergraduate teaching textbook with worked examples and questions throughout the volume. The discussion is generally of a very high quality and it is highly regarded as a standard reference book within the Paint Industry. Most of the subjects covered in this chapter are dealt with in greater detail in this book, however, application processes *per se* occupy only scant discussion when compared with the other areas. Viscoelasticity, an important area receives no treatment at all. This area of study is becoming increasingly important for coatings science.

Contained in the above book is a very detailed section dealing with specific paint and coatings industry methods for measuring rheology. Thus many of the industry standards are included and well described. However, for a fuller description of more exacting and non-paint industry specific methods, techniques, and equipment the book 'Rheological Measurement' by A. A. Collyer and D. W. Clegg, Elsevier, Amsterdam, 1988, is recommended.

At this level of reading it becomes necessary to understand an additional large body of non-Newtonian rheology, generally described as viscoelasticity. Briefly, this is a description of the coexistence of the viscous flow process already described and elastically stored energy processes (Hookes law). This area can be interesting, challenging, and often necessary to work with in order to make progress. Some of the general theories are best described in the book by G. Harrison, 'The Dynamic Properties of Supercooled Liquids', Academic Press, London and New York, 1976.

It is a very well written and readable text, but assumes some mathematical ability. Beyond this experimental treatment the field of rheology appears to diversify somewhat and is often dealt with as polymer rheology, for which I would recommend reading J. D. Ferry's, 'Viscoelastic Properties of Polymers', John Wiley and Son, Chichester, 3rd edn., 1980, or as colloidal rheology which is adequately dealt with in an introductory manner by J. W. Goodwin, Chapter 8 in 'Colloidal Dispersions', Special Publication No. 43, ed. J. W. Goodwin, Royal Society of Chemistry, London, 1982. Beyond these texts the specialist literature must be consulted.

Chapter 5

The Physics of Film Formation

A. B. PORT

INTRODUCTION

One of the most important steps in any coating process is film formation. This is the conversion of the coating from a liquid into a solid after application so that the coated article can start to be usefully employed. An obvious example from most peoples experience would be the painting of a domestic window and frame. The first stage of film formation is completed when the window can be touched without the decorator becoming painted. The coating is said to be 'touch dry' in this condition. Should the window be closed against its frame at this stage then it is likely that they would become stuck together under the pressure of closure.

The process of film formation must proceed further to avoid this problem and for the coating to behave more like a solid. In such a condition the coating is said to be 'hard dry' or block resistant under those particular conditions of time, temperature, and pressure. For a window manufacturer wishing to stack frames in a hot warehouse, further advancement of film formation would be necessary before unwelcome adhesion or blocking could be avoided.

It is now not quite so obvious what we mean by film formation as a coating becomes 'solid'. It is clear, however, that the rate and extent of film formation is of the highest importance both in terms of the process—how quickly a coated article can be handled, overcoated, or stored, and the coating performance—what properties have developed by what time after application.

The rationale for film formation behaviour in coatings is often explained in terms of glass transition temperature (T_g) and its relationship to viscosity (η) given by the Williams–Landel–Ferry (WLF) equation (equation 1):

$$\ln\eta_T = 27.6 - \frac{40.2(T - T_g)}{51.6 + (T - T_g)} \tag{1}$$

Here the viscosity of a polymeric glass is assumed to be 10^{12} Pa s, hence 27.6 from the value of ln 10^{12}. Using this expression and assumed values for viscosity at 'touch dry' or specific 'blocking' conditions the required T_g of the coating can be estimated relative to the test temperature, T. Whilst this approach gives useful guidelines it must be appreciated that the physical properties of a polymer are strongly dependent on temperature, elapsed time (from the start of the measurement), and the frequency of the measurement. This means that the value of T_g depends on the method of measurement and similarly the response of the polymer depends on test conditions. The dependence is especially strong when the polymer under examination is in its glass transition region. For film formation of coatings, if we attempt to describe the process by the indentation or flow under stress the position becomes more complex by virtue of the non-linear relationship between creep rates and stress. In summary the coating will behave more like an elastic solid than a viscous liquid if:

(i) the time of application of load is short,
(ii) the frequency of loading is high,
(iii) the magnitude of the applied stress is low,
(iv) the test temperature is low.

It is implicit in the description above that film formation depends on the coating's T_g increasing to approach or exceed the temperature under consideration. In the next section we will look at the ways in which this is achieved. There is, of course, a small but significant number of coatings where T_g is substantially below ambient, *i.e.* they are elastomeric in nature, and these will also be considered.

THERMOPLASTIC COATINGS

A traditional method of film formation used by the coatings industries for many years was simply to dissolve a polymer in a volatile solvent, apply it to the work piece, and allow the solvent to evaporate. For a number of reasons this approach has been superseded by others, described in later sections, and relatively few coatings technologies rely on simple solvent evaporation. The explanation for this loss in importance lies in the fact that many important polymer properties are achieved only when molecular weights become large enough for chain entanglement to occur. The values of this entanglement molecular weight, MW, will vary significantly for different polymers and range from a few thousand to in excess of 10^5 g mol^{-1}. In consequence, the viscosity of solutions of these polymers is relatively high and to maintain application viscosities the concentration of polymer has to be low (see Chapter 7). In turn this means that many applications are necessary to build up the thick coatings necessary for some end uses. The environmental legislative pressure to reduce solvent emissions, described earlier, has meant that such low solids coatings are no longer acceptable in many industries.

Film formation from solutions of thermoplastics occurs via the increase in T_g with loss of solvent. Initially, loss of solvent from these systems depends only on the vapour pressure of the solvent, how quickly solvent vapour is removed from the immediate environment, and the ratio of surface area to volume of the coating. As the system T_g increases, the rate will become dependent on the rate at which the solvent can diffuse through the solvent swollen polymer.

Solvent removal from the free surface sets up a solvent concentration gradient across the coatings to act as the driving force for diffusion. The normal treatments for diffusion of small molecules through polymers are made more complex for the coating systems where final T_g is above film formation temperature. In these systems vitrification will occur at some stage when the nominal T_g equals the drying temperature. Rates of diffusion of solvents in the glassy state are significantly lower than in the rubbery phase which means that glassy coatings can retain significant amounts of solvent for long periods—in some cases for a number of years. For some end uses this solvent retention is unacceptable, for example when coatings are in contact with food or drink, such as in can lacquers. Here extraction of solvents into the contents of the can would affect the taste and probably present a health hazard. In exterior durable paints the trace amounts of ethers, ketones, or halocarbons can act as initiators or 'fuels' for the photo-oxidative reactions causing weathering of finishes.

A further consequence of the evaporation of large volumes of solvents during film formation involving vitrification is that large shrinkage stresses can be set up. These internal stresses cannot be easily relieved in thin films adhering well to rigid substrates, and will weaken the coating's resistance to applied stresses or strains as well as the strength of adhesion to its substrate. To overcome the problems of solvent retention described above, stoving at temperatures considerably in excess of the coating T_g is carried out. This will not however remove internal stresses from the system when coating and substrate have substantially different coefficients of thermal expansion.

There are further limitations to thermoplastic coatings over and above environmental considerations especially where thermal or chemical resistance is required. To some extent these can be overcome using crystallisable polymers, but these impose their own difficulties in terms of forming stable solutions as paints. For solvent-borne coatings the more effective route to environmental compliance and higher levels of performance has been through the use of crosslinking polymers.

SOLUTIONS OF CROSSLINKING POLYMERS

The Crosslinking Process

Film formation from solutions of crosslinking polymer systems combines two processes. Solvent evaporation takes place as described for solutions of thermoplastic polymers. At the same time chemical reactions take place between functional groups on polymeric and/or monomeric species which result in the build up of molecular weight of the system. There is a wide diversity of polymer chemistries which are used to produce crosslinking systems and these are described in Chapter 7. The common feature however is that the average functionality of the component(s) must be greater than two. In a typical system, consisting of an A functional polymer or oligomer with a B functional crosslinker, reactions between A and B groups will lead to chain extension and branching (Figure 5.1). At some stage of the reaction macromolecules link to form a network of infinite molecular weight which is known as the gel. The process is termed gelation and occurs at a particular extent of reaction dependent on the

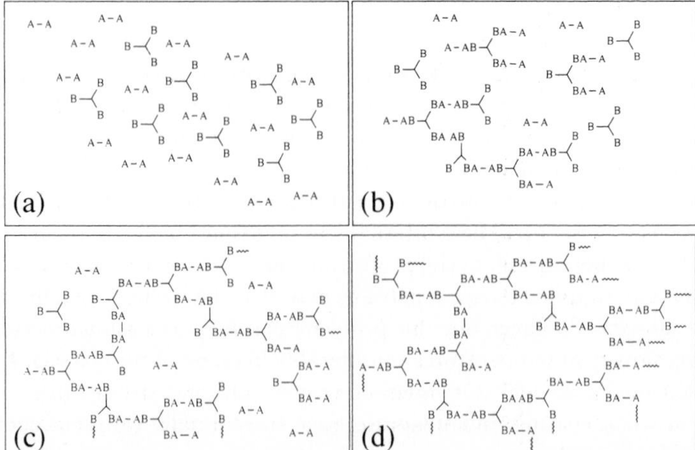

Figure 5.1 *Formation of a network of di-A- and tri-B-functional species.* (a) *Before reaction;* (b) *forming branched chains;* (c) *on gelation;* (d) *when complete*

ratio of the numbers of fuctional groups (stoicheiometry) and the effective functionality of the components. The relationship between conversion, stoicheiometry, and functionality for the system above is given by equation 2, where P_{gel} is the extent of reaction of the minor component at gelation, r is the ratio of the number of functional groups of types A and B, and $f_{e_{A,B}}$ the site average or effective functionality of A and B functional components. Where a mixture of A or B poly functional components is used it is important to use the site average rather than a simple number average functionality. Site average is defined by equation 3, where n_i is the number of moles of a component of functionality f_i. In comparison, the number average functionality, which has often been used incorrectly for f_e in equation 2, is defined by equation 4.

$$P_{gel} = \frac{1}{\sqrt{r(f_{e_A} - 1)(f_{e_B} - 1)}} \tag{2}$$

$$f_e = \sum_i \left[\frac{n_i f_i}{\sum_i (n_i f_i)} \right] f_i \tag{3}$$

$$f_n = \sum_i \left[\frac{n_i}{\sum_i n_i} \right] f_i \tag{4}$$

The distinction between f_e and f_n and its significance in gelation behaviour is illustrated in the examples in Scheme 1, for the simple polyesterification reaction between a diacid and a mixture of polyols at stoicheiometric ratio $r = 1$.

For single, monodisperse (in functionality) components, $f_e = f_n$ but for mixtures of components with different functionalities $f_e > f_n$. This reflects the higher probability of selecting a functional group from higher functional species in a polydisperse mixture. In network forming reactions this means that the higher functional species are incorporated more quickly and gelation occurs earlier in the reaction than if the number average was used. This feature becomes increasingly important when a polyfunctional ($f > 2$) film forming polymer is selected for a paint formulation. Whereas its simple number average functionality from the relationship $f_n = M_n$/e.w., where e.w. is equivalent weight, may be moderate, its effective functionality will be

For the following reaction mixture:
No. moles diacid = 1.00
No. moles diol = 0.50
No. moles tetrol = 0.25

$$f_{e_A} = \frac{\sum_i n_i f_i^2}{\sum_i n_i f_i} = \frac{1 \times 2^2}{1 \times 2} = 2 \qquad f_{n_A} = \frac{\sum_i n_i f_i}{\sum_i n_i} = \frac{1 \times 2}{1} = 2$$

$$f_{e_B} = \frac{\sum_i n_i f_i^2}{\sum_i n_i f_i} = \frac{(0.5 \times 2^2) + (0.25 \times 4^2)}{(0.5 \times 2) + (0.25 \times 4)} = \frac{2 + 4}{2} = 3$$

$$f_{n_B} = \frac{\sum_i n_i f_i}{\sum_i n_i} = \frac{(0.5 \times 2) + (0.25 \times 4)}{0.5 + 0.25} = \frac{2}{0.75} = 2.67$$

Then

$$P_{gel} = \frac{1}{\sqrt{r(f_{e_A} - 1)(f_{e_B} - 1)}} = \frac{1}{\sqrt{1 \times 1 \times 2}} \times 100 = 70.7\%$$

Using f_n

$$P_{gel} = \frac{1}{\sqrt{r(f_{n_A} - 1)(f_{n_B} - 1)}} = \frac{1}{\sqrt{1 \times 1 \times 1.67}} \times 100 = 77.4\%$$

Scheme 1

weighted in favour of higher functional species present as a consequence of the way it is made. In some cases $f_e \gg f_n$ and the gelation behaviour will be very different from that predicted by f_n alone.

The onset of gelation brings about very major changes in the physical form of a crosslinking system. Creation of an infinite molecular weight molecule causes M_w (the weight average MW) and viscosity to diverge. Processes such as flow and levelling which depend on a coating having a finite viscosity are essentially brought to an abrupt end at gelation.

Crosslinking reactions do not however cease at gelation. There is still a very large number of molecules and functional groups present at gelation in addition to the gel. These finite sized species are known as the sol since they are soluble and capable of extraction by solvent. Reactions continue and the proportion of sol steadily decreases whilst that of gel increases. For stoicheiometric mixtures capable of full reaction, sol fraction will decrease to zero and the gel fraction increase to one. Off-stoicheiometric mixtures or incomplete reaction will result in varying proportions of sol and gel. The characteristics of crosslinking systems can be summarised in Figure 5.2.

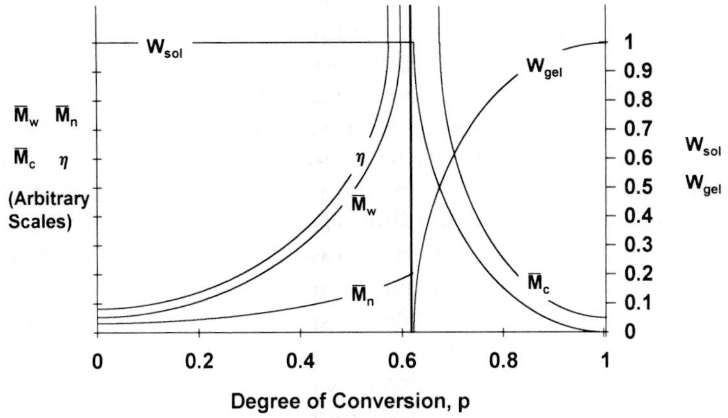

Figure 5.2 *The properties of a crosslinking system as a function of conversion in terms of MW averages, melt viscosity, and weight fractions of sol and gel*

The traditional definition of gelation is the point at which the elastic and viscous components of shear modulus intersect. In more practical terms it is marked by the rapid increase in viscosity and the appearance of a pronounced elastic solid behaviour from a previously viscous liquid. These changes provide the bases for empirical means of determining gel times or conversions such as 'snap time' and the point at which bubbles cease to rise in a reacting system.

Film formation in crosslinking systems might be thought to be complete at gelation since in the description above viscosity diverges at this point. Again we must refer to the arguments put forward at the beginning of this chapter and qualify the conditions under which we characterise the film. At gelation there will be a large proportion of low MW sol material and a solvent content which depends on the kinetics of crosslinking and speed of evaporation of solvent. Typically the system T_g will be below the test temperature and the material will behave as a weak, soft, or swollen rubber and may well not resist the indentation and blocking tests described earlier. Further reaction and loss of solvent will then be necessary before defined criteria of film formation are met.

Kinetic Aspects of the Crosslinking Process

The preceding section described the changes occurring during the crosslinking process largely in terms of extent of reaction. We must now address the rate at which these reactions occur and whether the physical state of the system influences the rate, especially when the system T_g approaches or exceeds the cure temperature. There are important reasons why the coatings scientist or technologist needs to understand and control the kinetics of the crosslinking process. Many coatings performance parameters depend critically on the extent of cure which is achieved in various timescales (see Chapter 6). In some industries the conditions of time and temperature for curing of crosslinking coatings are tightly specified. The reasons for this may be economic—as dictated by production line speed, size, and rating of ovens, or by the sensitivity of the substrate to elevated temperatures—or simply the traditional practice of an established process. Any paint manufacturer hoping to introduce a new product in these circumstances would have to offer significant improvements in cost or performance or both to induce the customer to make major changes in his operation. Thus, it is more usual for the coatings formulator to tailor his product to an existing cure schedule.

In contrast, there are examples where curing conditions are not subject to the same control. Coating products applied to large structures outdoors must achieve acceptable levels of film formation and performance whether in winter or summer and high or low humidity within production time scales which probably do not change significantly with the season.

A third aspect is that of the stability of the product before application. The customer or user of paint would like indefinite stability in the can, irrespective of storage conditions as well as fast complete cure after application. For ambient temperature crosslinking systems involving the simple reaction of functional groups on polymer and curing agent, the constraints of storage and reactivity are virtually impossible to overcome. Successful strategies have been to separate the reactive functionalised components into two or even three packs which are immediately mixed prior to use. Alternatively, a crosslinking chemistry may be employed for one pot systems which become operative only by the action of an atmospheric component such as oxygen, water, or more recently a vaporised catalyst. A number of examples, of this type are described in more detail in Chapter 7.

For stoving systems, thermally activated crosslinking chemistries mean that one-pack systems are more common. Changes in manufacturing practice towards lower stoving temperatures and increasing use of plastics materials, whilst maintaining or increasing production line speeds, all conspire to make the job of the coatings scientist more challenging.

Chemical Reactivity Control

In crosslinking systems where molecular mobility and functional group concentrations are high then for the reaction of Scheme 2 the overall rate of reaction depends on the conversion of an activated complex to product with rate constant k_2, and is independent of the size and mobility of reactants. In this case the temperature dependence of the rate constant is described by the Arrhenius equation (equation 5) where A is the pre-exponential constant for the reaction, E is the activation energy, R the gas constant, and T absolute temperature. In the early stages of the crosslinking reaction the Arrhenius equation provides an adequate description of the system. When combined with knowledge or assumption of the reaction mechanism and definitions of acceptable stability and reactivity, then approximate guidelines for control and design of crosslinking kinetics can be obtained.

$$\sim\!\mathrm{A} + \mathrm{B}\!\sim \underset{k_{-1}}{\overset{k_1}{\rightleftharpoons}} \sim\![\mathrm{AB}]^*\!\sim \xrightarrow{k_2} \sim\!\mathrm{AB}\!\sim$$

Scheme 2

$$k_2 = A \exp\left(-\frac{E}{RT}\right) \tag{5}$$

The limitation of this treatment is that as the crosslinking reaction proceeds MW builds rapidly and the system T_g will increase. When T_g climbs to within approximately 50 to 100 °C of cure temperature then k_1,k_{-1} become comparable with k_2 and diffusion control operates.

Diffusion Control

The overall rate of reaction now depends on how quickly the components carrying the functional groups can move through the reaction medium to allow reaction to take place. In these circumstances, the free volume relationship of the modified WLF equation (equation 6) is likely to give a better description of the temperature dependence of the rate constant where C_g is the so called 'universal constant' in the WLF equation and is approximately 51 °C.

$$\ln K \propto \frac{T - T_g}{T - T_g + C_g} \tag{6}$$

The changeover from activation to diffusion control is demonstrated in data reported by Dusek for an epoxy-amine system (Figure 5.3a). Epoxy conversion with time was measured at a number of different temperatures. Initially the conversion curves could be made to coincide by shifting along the time axis using the factor A indicating activity control. With increasing extent of reaction or conversion, T_g increases as seen in Figure 5.3b, and deviation from the master curve occurs as T_g approaches cure temperature indicating diffusion control.

At high extents of reaction in crosslinking systems, the concentration of remaining functional groups will be low and increasingly likely to be 'fixed' on the gel. Thus the final stages of reaction may be exceedingly slow or (in practical timescales) will never occur. This is an example of specific diffusion control and often termed the topological limit of crosslinking reactions. There is evidence to suggest that for some epoxy systems this limit is about 90–95% conversion.

The WLF expression (equation 6) above suggests that crosslinking reactions will continue until $(T - T_g) \approx -50$ °C, *i.e.* the system T_g is 50 °C above the cure temperature. In practice, the rate of reaction slows very appreciably in the region where T_g equals cure temperature when the system is said to undergo vitrification, *i.e.* become glass-like. For ambient cured coatings, the practical limit of $T - T_g$ is often approximately -25 °C to -35 °C or system T_gs of 50 °C to 60 °C where ambient is 25 °C. By exercising choice of solvents with different volatilities, the system T_g may be depressed for long enough (by solvent plasticisation) for further reaction to occur so that when the solvent does evaporate the final T_g may reach higher temperatures in some examples up to 90–100 °C.

Thus, it can be recognised that the design of coating systems to achieve specified performance related conversions under varying cure conditions is complex. It requires knowledge of conversion–performance and T_g–conversion relationships as well as reaction kinetics and solvent evaporation behaviour under different application conditions.

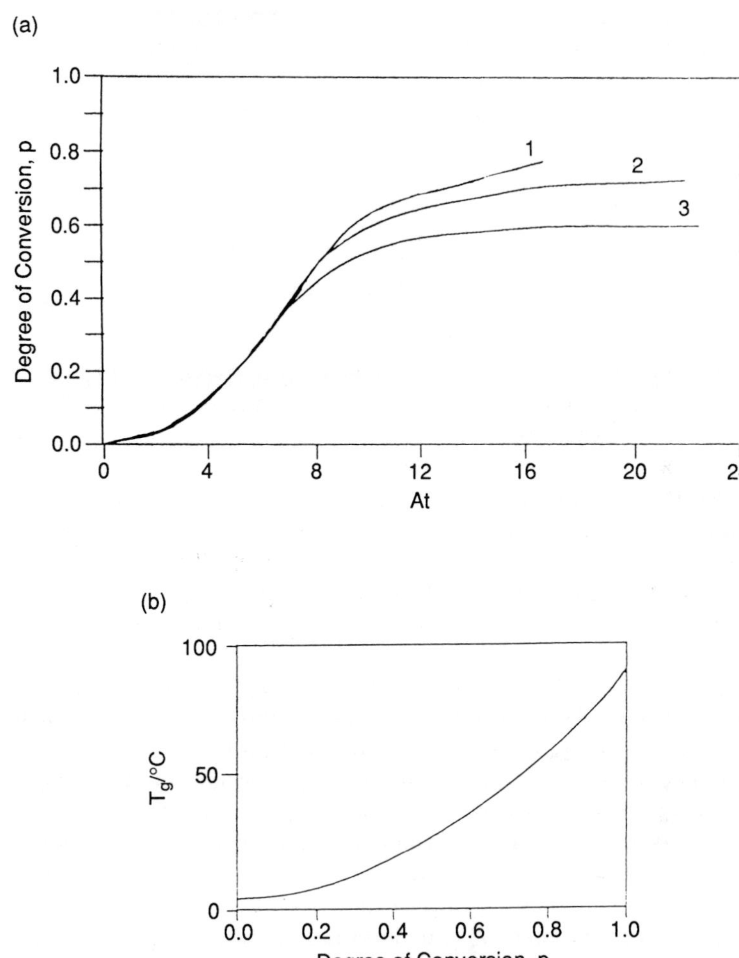

Figure 5.3 (a) *The onset of diffusion control in epoxy-amine network formation. Reaction temperature* (°C): 1 = 100; 2 = 64; 3 = 40. A *is a temperature dependent multiplicative factor.* (b) *Dependence of* T_g *on extent of conversion* (Reproduced with permission from K. Dusek, *Adv. Polym. Sci.*, 1986, **78**, 1.)

T_g–Conversion Relationships

For solvent free systems some useful information can be gained from isothermal curing studies where the data is presented in terms of a time–temperature–transformation diagram developed by Gillham and co-workers. In such diagrams (see Figure 5.4) the conditions, times, and

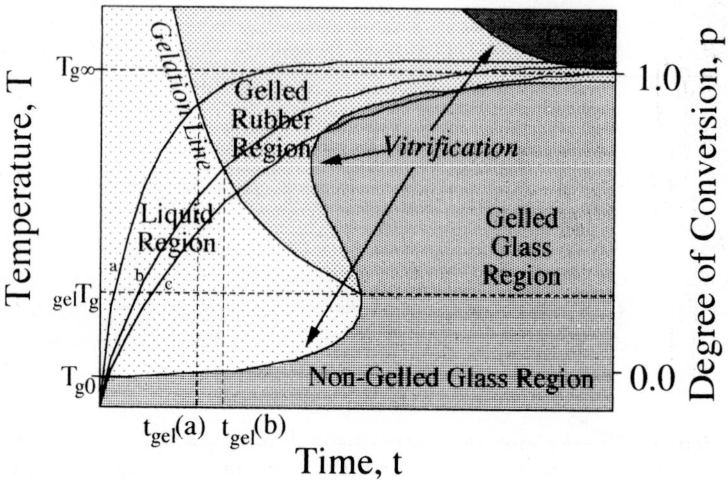

Figure 5.4 *Time–Temperature–Transformation phase relationships for a thermoset polymer*
(Reproduced with permission from J. B. Enns and J. K. Gillham, *J. Appl. Polym. Sci.*, 1983, **28**, 2567.)

temperature to gelation and vitrification are presented in a type of phase diagram. If the relationship between T_g and conversion is known then time–temperature–conversion curves can be drawn.

From Figure 5.4 it can be seen that there will be a maximum T_g for a particular system ($T_{g\infty}$) and that it is necessary to cure at or above this temperature to achieve it. The value of $T_{g\infty}$ is dependent on crosslink density and the maximum achievable extent of conversion. Intuitively, we would expect that as the distance between branch points or network junctions decreases then T_g will increase. Nielsen has described an empirical relationship (equation 7) which relates the rise in T_g (ΔT_g) with (full) conversion to network with M_c, the number average molar mass between elastically effective network junctions.

$$\Delta T_g = \frac{3.9 \times 10^4}{M_c} \tag{7}$$

For highly crosslinked systems, such as epoxies and phenolics, M_c is typically 300 to 1000 giving ΔT_g values of 130 °C to 40 °C. For moderately crosslinked polymers, such as those based on thermosetting polyesters, M_c is typically 1500 to 5000 leading to ΔT_g of 25 °C to 8 °C.

The magnitude of changes in T_g on crosslinking makes it a useful means of monitoring or following the process. Recently, theoretical treatments have successfully predicted the non-linear relationship between T_g and extent of conversion.

SOLVENTLESS CROSSLINKING SYSTEMS

The most important class of coatings in this category are powder paints. Their technology is described elsewhere in this book and, of course, the important characteristics of crosslinking systems described above apply to powder coatings. Film formation in electrostatic spray powder systems starts with a solid in particulate form on the workpiece. On stoving, the individual particles soften and flow and coalescence takes place. During this early stage in film formation it is hoped that complete wetting of the substrate occurs, entrapped air is released, and an even film of now liquid coating covers the surface. Simultaneously, however, the crosslinking chemistry starts to work giving rise to MW and viscosity build as described earlier. Near gelation, flow and levelling cease and the surface appearance at that point is fixed in place.

The viscosity and temperature profile during stoving might typically be as shown in Figure 5.5. The so called 'viscosity well' describes how fluid the system becomes and for how long. It is determined by a number of

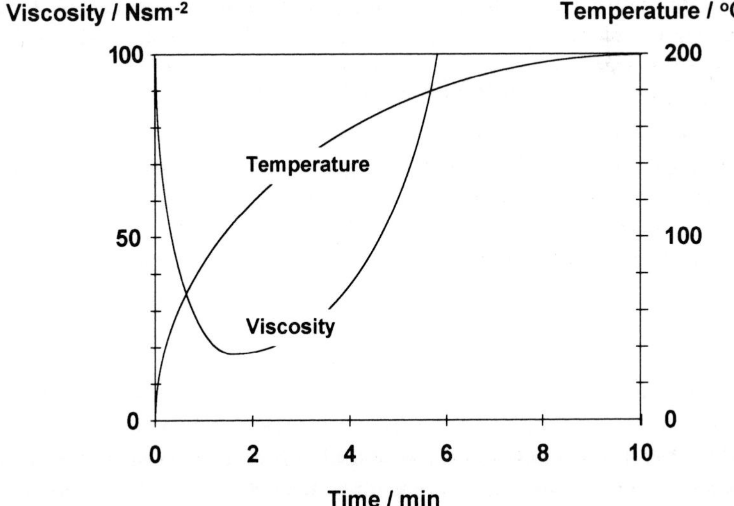

Figure 5.5 *The viscosity and temperature profiles for a powder coating during stoving*

factors including rate of temperature rise, crosslinking reaction kinetics, MW build characteristics prior to gelation, and extent of reaction at gelation. This is in addition to MW, viscosity, and pigment volume concentration of the starting materials. The normal 'orange peel' appearance of most powder coatings would indicate that there is still room for improvement in balancing the rheological and reactivity characteristics of these systems.

Powder coatings are one pack systems where it is hoped each particle contains all formulation ingredients in the right proportion. Storage prior to use places requirements both on physical and chemical stability. To avoid physical sintering the powder, T_g needs to be above about 50 °C again depending on storage temperature and hydrostatic pressure. Chemical stability in the solid state does not usually pose too many problems since the raw materials are in a glassy state when reaction rates will be very slow. Traditional methods of manufacturing powder coatings however require the use of extruders to mix and disperse the ingredients into the polymer matrix. Reaction will inevitably occur in the extruder operating at typically 100–120 °C. Such pre-reaction will at best degrade the flow and levelling and hence appearance of the final coating. The problem is exacerbated by the demand for powder coatings which cure at ever lower temperatures for use on temperature sensitive substrates. Currently the lower limit on stoving temperatures is about 120–140 °C.

Important solventless-liquid systems include radiation cured and two pack chemistries, see Chapter 7. They form films more readily than powders, but are subject to a varying extent to diffusion control.

DISPERSE PHASE SYSTEMS

The problems of using solutions of high MW thermoplastics in coatings have been successfully overcome by using disperse phase technology. This approach is described in more detail in Chapter 8. The advantages of stabilised dispersions of polymer particles in an aqueous continuous phase are those of low viscosity, low VOC, and the possibility of using very high MW polymers. To take advantage of the properties offered by such high MW, the process of film formation in these systems is of paramount importance.

After application of these coatings water and organic co-solvents begin to be lost by evaporation, the solids contents rise, and particles approach each other more closely. For successful film formation, the particles must coalesce and this involves overcoming the hitherto effective repulsive interparticle stabilisation forces. The driving forces for coalescence are thought to arise from capillary forces as particles get very close and/or the surface energy reduction as the sub-micron particles form a continuous film. Whatever the origin the forces are also such that substantial deformation of the spherical particles into closer packed polyhedra occurs in the latter stages of film formation.

Film formation requires diffusion of polymer chains across the particle interface which, for ambient systems, means that the T_g of the polymer in the interfacial regions is below ambient. This does not necessarily mean that the particle 'bulk' T_g has to be low. Organic co-solvents or coalescing aids will plasticise the polymer for long enough to allow film formation and when lost by evaporation the coating T_g will climb to above ambient. An alternative is to create multilayer or gradient composition particles through controlled emulsion polymerisation techniques. Here it is possible to create a high T_g core with a low T_g shell in a particle to combine high T_g properties with low film formation temperatures.

The advantages of crosslinking the coating are difficult to achieve in ambient disperse phase systems. Such systems are beginning to appear on the market but are not yet in widespread use. For stoving systems however a number of crosslinking chemistries have been available for some time finding widespread application in markets such as packaging and coil coatings. Film formation and crosslinking may have to take place extremely rapidly in such end uses where stoving schedules may be less than a minute with metal temperatures rising to 250 °C or higher. Subsequent performance in hot aqueous or corrosive environments relies on the quality of film formation by coalescence and the absence of weak boundary pathways for the ingress or egress of corrosion reactants/products.

BIBLIOGRAPHY

Z. W. Wicks Jnr., F. N. Jones, and P. S. Pappas, 'Organic Coatings Science and Technology—Film Formation, Components, and Appearance', Vol. 1, John Wiley and Sons, Chichester, 1992.

Chapter 6

Performance Properties of Coatings

A. B. PORT

INTRODUCTION

The performance of a coating describes how well it is carrying out its function in service. It will be appreciated from other chapters that coatings are required to fulfil a broad range of roles and provide a number of effects. This in turn leads to diverse classifications for coatings performance and it would be beyond the scope of this book to attempt to describe in detail all types of performance, the coatings technologies to achieve them, and the procedures used to predict or monitor those performances.

The approach which has been taken in this chapter is to concentrate on coatings performance in terms of its response to physical, mechanical, and chemical abuse. In particular, the maintenance of film integrity under these forms of attack is chosen, the rationale being that once the coating is cracked and/or lost from the substrate, its primary roles have been at least compromised. The retention of this performance in service will be an important factor in determining the lifetime of a coating system and so the influence of chemical and physical ageing is also addressed.

MECHANICAL PERFORMANCE

The mechanical performance of a coating describes how it responds to stresses and strains imposed on it during service. This performance is rarely the sole criterion, rather it is one of a group of properties which must be achieved. In this way the importance of mechanical properties may vary from being a key factor to being one of secondary consideration. Nevertheless, failure to resist some form of physical or mechanical abuse, which results in cracking, and any subsequent loss of coating must always be regarded as a limiting performance parameter.

The severity and frequency of the mechanical abuse clearly plays an important part in determining the relative position of mechanical performance both in the hierarchy of properties to be met and also at which stage it is considered during product development. By way of an example, coatings intended for some can and coil end uses have been designed around flexible and extensible polymers because of the severe deformations involved in coil forming and the manufacture of two piece cans by the draw–redraw (DRD) process. In comparison, when mechanical performance is of secondary importance, a coating's mechanical properties may be addressed by the formulator at a relatively late stage of product development. The opportunities for such 'last minute' formulation changes are becoming fewer with the increasing constraints imposed by environmental legislation. It is now increasingly important for modifications to the formulation to avoid contributing to the VOC level. This factor obviously limits the choices available and the traditional approaches of adding fillers, extenders, plasticisers, co-film formers, *etc.* cannot be used where the penalty of higher viscosity has to be offset by additional solvent.

For this reason, and others, it becomes more important to try to target the balance of properties including mechanical performance through design of the system from the start of the coatings development. In order to do this the key links between coating performance, material properties, and coating/polymer design have to be established. Given the complexity of each of these subjects it is not surprising that there are few quantitative performance–property–structure relationships in coatings science. The potential benefits, however, are such that this approach to coatings development is the subject of much effort in the coatings industries. If successful it should be possible to learn important lessons from each coating product development and in such a way that the knowledge gained and the emerging technology is transferable within the various coatings business sectors.

Performance Specification

Physical and mechanical performance requirements of coatings in service vary enormously with the different end use applications. A number of examples will serve to illustrate the diversity and complexity of the applied stresses and strains.

Marine Coatings. A general purpose anti-corrosive and anti-abrasive paint might be required to perform equally well on a ship's side or deck or as a hold coating. On the side of the ship at sea the coating experiences both diurnal temperature fluctuations and intermittent water cooling/heating. The difference in thermal expansion between paint and metal results in cyclic compressive and tensile stresses in the coating. There may also be dilational and shrinkage stresses set up in the coating as water is absorbed or desorbed. The stresses and strains involved are relatively low and are applied at low rates. On berthing however, as the ship is unloaded and loaded it moves against the fenders at the dockside (Figure 6.1). Here moderate to high shear stresses are exerted on the coating at low strain rates as the ship rises and falls. Within the hold, the coating will sustain high speed impacts as a cargo falls from heights of up to 30 m. The cargo may be hard, angular particles so that local stresses and strain rates on the coating may be very high. On deck a complex mixture of stresses, strains, and strain rates will be expected as people and machinery go about their work but with the added complication of chemical attack from liquid cargoes, lubricants, *etc.*

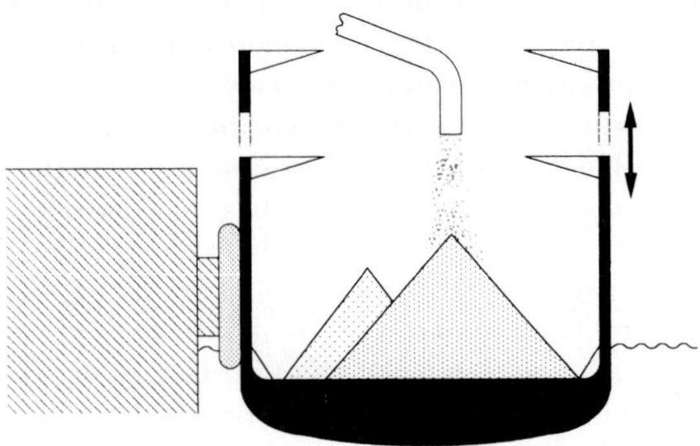

Figure 6.1 *Stresses and strains on marine coatings*

Can Coatings. In the draw–redraw process for making two piece cans, coated metal is drawn by punch and die into progressively taller, narrower can bodies. Strains in the direction of drawing may be in the order of 100% whereas strains normal to that direction are in the order of −50% (to maintain approximately constant wall thickness) (see Figure 6.2). Maximum punch speeds are a few metres per second so that strain rates will be relatively high. Metal and tooling temperatures will be above ambient due to the amount of work done on the metal during forming.

Further forming processes are carried out on the can body after drawing which may include putting expansion rings into the base,

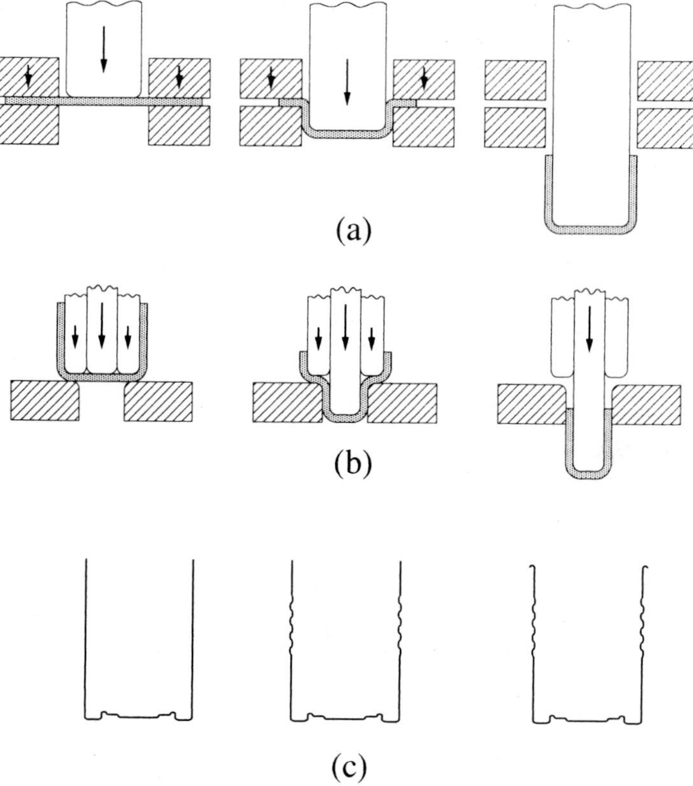

Figure 6.2 (a) *The drawing process,* (b) *the redrawing process, and* (c) *further forming processes, in can manufacture*

beading into the can wall, trimming the can to the correct height, and providing a flange for end seaming (Figure 6.2c).

Each of these processes occurs at the high speeds commensurate with modern can making. Following this somewhat traumatic process of manufacture, the coated can body is filled, typically with foodstuff, and processed for anything up to 90 min at 130 °C for storage prior to use by the consumer. The coating must continue to provide corrosion protection during processing, transport, and storage for many months. It is thus vital that these thin coatings (7–15 μm) retain their integrity and adhesion to the substrate during all of these processes.

Building Paints. In contrast to the two examples above it would seem that a paint intended for decoration and protection of exterior woodwork has few mechanical demands placed upon it. This substrate is, however, dimensionally unstable and expands and contracts with changes in temperature and, more importantly, with moisture ingress and egress. For a continuous substrate the stresses and strains involved are relatively small and usually occur over protracted timescales. When the wood cracks on drying, local strains may become very high. These strains must be recoverable should the wood take up water, swell, and close the cracks otherwise buckling or blistering of the coating will take place (Figure 6.3).

The retention of elasticity or plasticity under exposure to sunlight and moisture is clearly important here. Similarly, the maintenance of strong adhesion to this variable substrate is a key performance parameter.

Figure 6.3 *Distortion of a building paint coating on wood*

The coatings scientist must have a clear understanding of the nature of the mechanical demands imposed in a particular end use. The critical abuses or performance criteria need to be established and characterised in terms of the types and magnitudes of stress, strain, and strain rate, temperature, and frequency or timescale. Where possible, quantification of these parameters should be undertaken although it is clear from the previous examples that this is often difficult. The analysis should be attempted at some level so that the appropriate laboratory test protocol can be devised and relevant material properties measured. In some specialised applications a set of material properties, which have been found to relate to satisfactory performance in service, have been specified by the end user. Here any new product under development needs to achieve the specified properties before the customer would consider the coating for his application. Examples of this type would include paints for the automotive and aviation markets. This approach occasionally causes problems when a specification is written around a particular coating technology found to be successful in an application. A new coating might well achieve satisfactory performance levels in service but fail the specification tests because the tests do not accurately reflect what is required in the application, only the properties of the previous material. In such cases the arguments for the new coating must be sound and persuasive (or the customer must want to adopt the new technology) to overcome the inertia and cost implications by which an established system resists change.

The quality of the information gained on the service environment and a coating's performance in that environment can be a major contribution to the successful development of a coating and its introduction to the marketplace. In the absence of such quality information, much is left to the experience of the coatings formulator and product development is guided by comparisons with previous products or those of a competitor. In either case it is usual to employ a number of mechanical performance tests and material property determinations in coatings research and development and these are described in the next section.

Traditional Mechanical Performance Tests

Examination of the three examples above of coatings undergoing physical and mechanical duress suggests that the properties of hardness, flexibility, impact resistance, formability, adhesion, wear, and friction would be important parameters in describing mechanical performance. Laboratory tests have been devised to provide empirical or semi-quantitative measures of these properties of coatings, usually on the substrate of interest. Some of these tests have been sufficiently widely used to be covered by national standards, *e.g.* BS3900 in the UK. There are far more tests however which have been devised to emulate coatings in specific applications sometimes peculiar to particular industries or even individual companies. Clearly it would be beyond the scope of this chapter to describe all of these tests and the sections below deal with those perhaps most widely used.

Hardness. 'Pencil Hardness' is quoted as the grade of pencil which either marks the coating or causes a deep scratch in the surface. This test can be somewhat subjective and operator dependent. These problems are overcome by using the more classical technique of measuring the penetration of a known indenter under a given load. Care must be taken when using this method for thin films on rigid substrates. When the depth of penetration becomes comparable with film thickness the substrate influences the result making the coating apparently harder. Similar effects can be seen for multilayer coatings of different thicknesses. Pendulum hardness is also widely used in the coatings industry. Here a hard knife edge mounted in a pendulum is set rocking through a given angle on a coated panel, and the number of swings counted before a certain angular decay has occurred. It should be noted that the result depends not only on hardness in terms of the depth of penetration of the knife edge, but also on the damping properties of the coating, *i.e.* the amount of energy absorbed during cyclic deformation.

The most rapid, cheapest, and most subjective tests are the thumb nail and the pen knife. These are widely used in the coatings industries and despite their crude nature are still powerful in the hands of an experienced paint maker and formulator.

Flexibility. A number of tests have been devised to determine the severity of bending of a coated panel needed to cause failure in the coating as seen by cracking. The simplest of these is the T bend test where a coated panel is folded back on itself repeatedly until the coating remains intact (Figure 6.4). The most flexible coating will survive the first folding (zero-T), less flexible ones will be one-T, two-T, up to four-T.

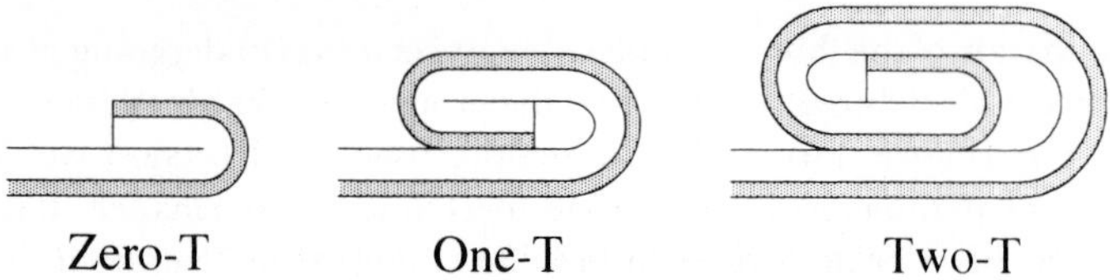

Figure 6.4 *T bend tests*

A series of hinges with different diameters at the apex can be used to fold panels. Again the most flexible panel will survive the most severe, lowest diameter bend. This is known as a wedge bend or single diameter mandrel test. An alternative to this is the conical mandrel where a panel is folded around a single mandrel in the form of a cone with diameters ranging from 3 to 37 mm. The diameter at which folding occurs without cracking is noted; again the most flexible coating will survive the tightest bend without cracking.

As will be described later, the properties of the coating depend on temperature, rate of deformation, film thickness, age, and conditioning so that care should be taken to conduct these tests consistently.

Impact Resistance. Falling weight impact testers are widely used to assess a coating's resistance to rapid loading and deformation. Typically a flyer or indenter of known mass is dropped through a known height onto a supported coated panel. In one arrangement, indenters with one hemispherical end are used on light gauge aluminium or steel panels such that the panel is substantially deformed. Combinations of weight and height are used to deliver increasing impact energies, expressed as Nm or J, until cracking and delamination are observed either on the front (forward) or back (reverse) of the panel. The panel material type and thickness should be specified since the amount of metal deformation will be affected, as will the impact energy for failure. For some end uses, heavy metal plate is used in which case deformation of the substrate is much less. Care should be taken when repeated tests at different energies are used on the same plate since the area of damage from previous impacts can extend considerably from the point of impact. This damage may not be immediately visible to the operator but ultrasonic tests (or the pen knife) will reveal polymer–substrate disruption or delamination which will compromise subsequent results if impacts are made too close to each other.

Abrasion and Wear. This is probably one of the most difficult laboratory tests to relate to in-service performance. A number of designs have been used including the measurement of coating weight loss under abrasive wheels, falling sand, or gravel ('gravelometer').

Formability. A number of test designs have been developed to emulate the forming, drawing, and stretch forming operations carried out on coated metal. In addition to the T bend test described above, box draw tooling and the Erichsen cupping press have been used. Again it is necessary to apply representative tool/substrate temperatures and deformation rates as well as lubrication to obtain data relevant to particular operations.

Friction. A simple apparatus to measure static and dynamic coefficients of friction under load involves pulling a weighted sledge across a coated panel. Where higher pressures are applied to coatings, such as occur during the movement of coated metal into the die in the DRD process described above, hard polished balls may be used. Choice and weight of lubricant film on the coating and condition of load bearing surfaces needs careful attention to avoid spurious results.

Adhesion. This important performance property of coatings remains an area of controversy. The pen knife and its slightly more sophisticated multi-bladed version known as the cross hatch tester remain the most widely used methods of assessing coating adhesion. There are few if any methods for quantifying the adhesive strength of thin coating films to their substrates. Some success in terms of method and analysis has been achieved in the blister test developed by Briscoe and others. Here air or fluid pressure is applied behind the coating to grow a blister (Figure 6.5a). The so-called bolt-pull test, where a bolt or similar is bonded to the coating and the load to cause coating–substrate debonding is measured (Figure 6.5b), is irreproducible and analytically flawed. Similarly, lap shear testing is problematical and irreproducible. The Peel test (Figure 6.5c), is unsuitable for most glassy coating systems.

It is probably fair to say that the extremes of good and poor adhesion present no problems to the coatings scientist. At one extreme any attempts to remove the coating result primarily in cohesive failure of the coating itself, whilst at the other coating adhesion failure is only too evident. When

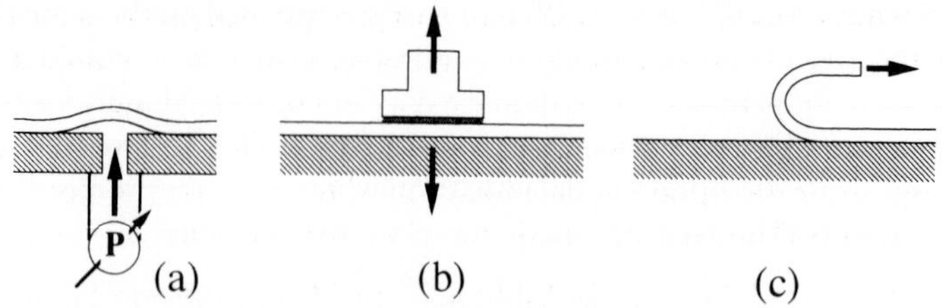

Figure 6.5 (a) *The blister test,* (b) *the bolt-pull test, and* (c) *the Peel test, for adhesion*

adhesion is marginal, a sensitive quantitative method of measuring the strengths of adhesion at the interface with its substrate (metal or another coating) is still required. Without such a method polymer design and coating formulation for improved adhesion become very difficult especially in view of the complexity and importance of surface preparation, application conditions, and environmental variability.

Material Properties of Coatings

In addition to the traditional paint tests, increasing effort has been put into determining more fundamental materials properties of coatings. This normally requires the preparation of free films of the coatings or, in some cases, 'bulk' specimens a few millimetres in thickness. Free films are conveniently prepared by applying the coating by an appropriate method to low energy surfaces such as fluoropolymer sheet, fluoropolymer coated metal, or substrates treated with release agents. Bulk specimens will usually require tooling for casting or compression moulding. This is relatively straightforward for zero VOC systems which evolve no volatiles during cure or film formation. Where solvent, water, or other volatile condensation products are generated, they must be allowed to escape otherwise voided specimens will result.

Once the difficulties of specimen preparation are overcome, a powerful range of materials property measurement and characterisation techniques is accessible. These techniques can give valuable quantitative information on the stress–strain behaviour, dynamic and thermomechanical properties, and the deformation, yield, and fracture response of the coatings. Time dependent or viscoelastic properties can be measured and can be used to predict behaviour at extremes of time and temperature not conveniently available in the laboratory. Much of the data produced in these experiments can be directly related to the large body of published information on polymeric materials properties. The theory and analytical treatment developed elsewhere can be applied to these data to characterise the structure of the coatings and to suggest the micro and macromolecular origins of the macroscopic response of the material to stress and strain. The specimens themselves can be used in other experiments for the characterisation of structure and morphology of the materials and so aid the development of structure–property relationships. Examples of these techniques would include scanning and transmission electron microscopy for examination of morphology and fracture surfaces, scattering techniques for identification of fine structure and ordering, and solvent swelling methods for network characterisation.

Stress–Strain Properties. In the simplest and most commonly used stress–strain test a specimen is deformed at constant rate and the resultant stress measured by some form of load cell (Figure 6.6a). Results are plotted as stress (load/original cross sectional area of sample) against strain (extension/original length of specimen). Polymers and, hence, organic coatings exhibit a wide range of stress–strain behaviour from hard and brittle (A) through tough, ductile (B), to soft elastometric (C) (Figure 6.6b).

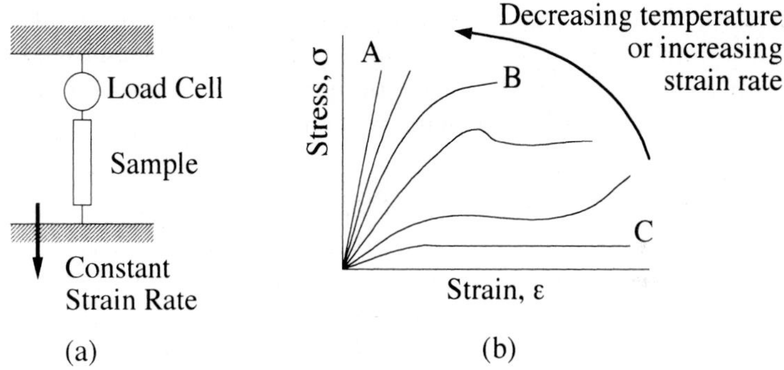

Figure 6.6 *(a) Stress–strain measurement and (b) typical representation of stress–strain results*

The form of the stress–strain curve is influenced not only by the structure of the polymer but also strongly by temperature and strain rate (as indicated in Figure 6.6b). This dependence is strongest when the coating T_g and test temperature are reasonably close say ± 30–$40\,^{\circ}$C. Residual solvent, water, or other plasticising species can cause significant shifts in T_g and hence the tensile stress–strain curves. Failure stresses and strains are also particularly difficult to measure for samples exhibiting brittle fracture. For these materials, failure is governed more by the flaws present and the material's ability to resist propagation of these flaws and cracks under stress.

In conclusion, although stress–strain tests are relatively quick and easy to do the interpretation of the results is more difficult and needs to be done carefully. However if the conditions of stress, strain, strain rate, and temperature for an end use are known then the generalised response of the coating under those conditions can be easily obtained through these tests. Similarly if the desired response appropriate to the conditions of abuse is known, appropriate classes of polymer technology can be selected from experience or published data.

Rubber Elasticity Theory. Many crosslinked polymers above their glass transition temperature exhibit elastomeric or rubbery elastic behaviour. Stress–strain measurements in this rubbery state can be treated by the classical theory of rubber elasticity to yield information on the network structure. Published data on so-called ideal networks can also be used to give an insight into the influence of network structure on physical and mechanical properties. In this context the distance between branch points, functionality of branch points, imperfections, and, by inference, extent of cure all have an influence on stress–strain properties in the rubbery state and, as will be described below, these effects may still operate in the glassy region when large strain properties are considered.

In its simplest treatment, measurements of shear (G) or tensile modulus (E) can be obtained from the slope of the initial stress–strain curve or from dynamic mechanical data [equations (1) and (2)]

$$G \cong \frac{\rho RT}{\bar{M}_c} \tag{1}$$

$$E \cong \frac{3\rho RT}{\bar{M}_c} \tag{2}$$

where R is the gas constant, T is absolute temperature, ρ is density, and $\bar{M}_c$ is the number average molecular mass between junctions. More rigorously, equilibrium stress (f) should be plotted against $(\alpha - \alpha^{-2})$ where α is the extension ratio of length/original length. Such plots usually give good straight lines from which ν, the number of moles of elastically effective network chains per unit volume, can be obtained [equation (3)]:

$$\frac{f}{(\alpha - \alpha^{-2})} = G = \frac{\rho RT}{\bar{M}_c} = \nu RT \tag{3}$$

Dynamic Mechanical Properties. Dynamic mechanical testing measures the response of a polymer to sinusoidal or other cyclic stress (or strain). For an ideal elastic material, stress and strain would be in phase, whereas an ideally viscous response has stress and strain 90° out of phase. Real polymers are viscoelastic, in other words possess both elastic and viscous components and strain lags stress by the phase angle δ. There are many designs of equipment used to generate dynamic mechanical data. The most useful of them are able to vary independently both the applied frequency of the cyclic stress and the temperature of the specimen. Depending on the geometry of the instrument, the stress may be applied in tension, shear, or oscillatory torsion between parallel plates. In general the results are expressed as a complex modulus, for example a complex Young's Modulus, E^* given by equation (4):

$$E^* = E' + iE'' \tag{4}$$

$$(\text{where } i = \sqrt{-1})$$

E' is the phase or storage or real part of the modulus, and E'' is the out of phase or loss or imaginary part of the modulus. The ratio of E''/E' gives the ratio of energy lost as heat to that stored during one cycle of deformation and equates to the tangent of the loss angle, δ in equation (5):

$$\tan \delta = \frac{E''}{E'} \tag{5}$$

Dynamic mechanical testing is a powerful technique for characterising polymers and coatings. It is one of the most sensitive techniques for determining primary and secondary transitions and for probing the structure of multiphase systems such as semi-crystalline polymers, polymer blends, and toughened systems. In addition to this, it has been used in coatings for determining extent of cure (from T_g and the shape and intensity of the tan δ peak at T_g), crosslink density (from the value of rubbery modulus), and determining the optimum stoicheiometry of polymer and crosslinker (from maximum T_g) (Figure 6.7).

Time Dependent Properties: Creep and Stress Relaxation Techniques. Creep is the time dependent deformation of a material under constant stress (Figure 6.8a), whereas the decay of stress at constant elongation is termed stress relaxation (Figure 6.8b). Data from these experiments are typically expressed as creep compliance curves (Figure 6.9a), where compliance is the inverse of modulus and relaxation modulus–time curves (Figure 6.9b).

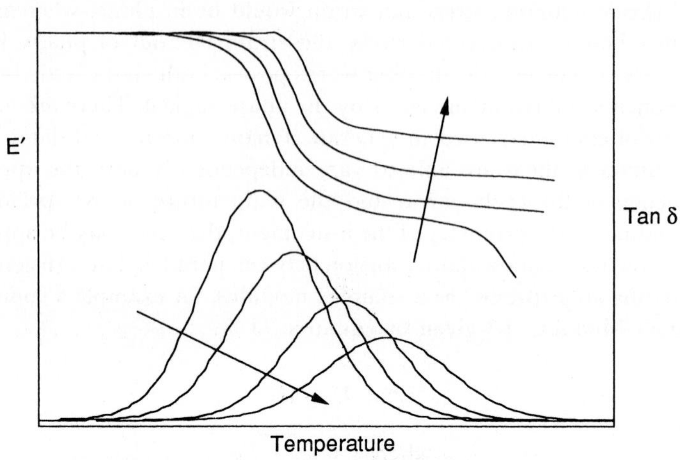

Figure 6.7 *Typical dynamic mechanical data (arrows indicate increasing extent of cure or optimisation of reaction stoicheiometry)*

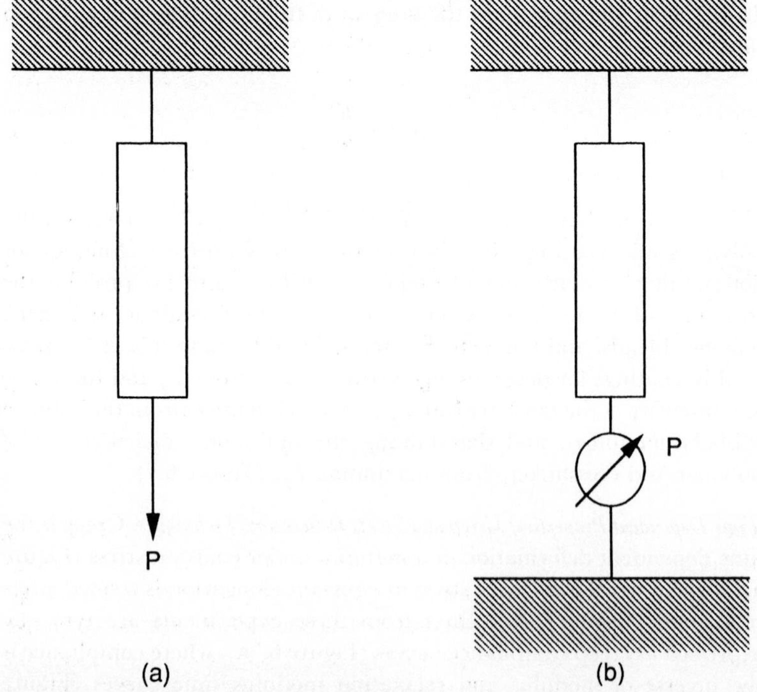

Figure 6.8 *Creep and stress relaxation experiments*

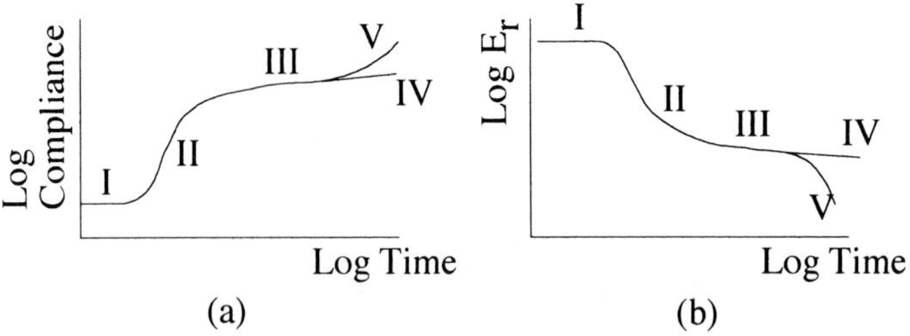

Figure 6.9 *Creep compliance and relaxation curves*

At short times, the material behaves as a hard elastic solid and rates of creep/strain rate are low (region I in Figure 6.9b). At longer times, rates of creep increase rapidly (region II) as the material gradually alters its response from hard to soft elastic behaviour (region III). At still longer times, an uncrosslinked polymer will behave as though it were a viscous liquid (region V) and large scale irrecoverable deformation takes place. When the polymer is crosslinked rates of creep/strain rate remain very low (region IV). The stabilisation of creep rates for non-crosslinked polymers in region III is due to entanglements acting as physical crosslinks and so the length of this plateau is MW dependent. For many polymers, the timescales necessary to observe these different regions are inaccessible at either extreme. To overcome this problem the apparent similarity between modulus–time (Figure 6.10) and modulus–temperature curves (Figure 6.7) is exploited in the time–temperature superposition principle. For example stress relaxation data collected over convenient timescales at different temperatures can be shifted along the log time axis relative to a particular temperature to yield a master curve (Figure 6.11). The individual curves are shifted horizontally until they form a smooth single curve. This can either be done 'by eye' or using the shift factor as predicted by the WLF equation (equation 6):

$$\log a_T = \frac{17.5(T - T_g)}{51 + (T - T_g)} \tag{6}$$

when the reference temperature is T_g, other numerical constraints must be used if the reference temperature is other than T_g. This effectively describes a material's response over many decades of time or frequency.

These techniques have been used to address a number of performance issues for coatings, for example the rapid assessment of block resistance and cold flow of architectural paints under pressure and the physical stability of powder paints at high storage temperatures. For abrasion

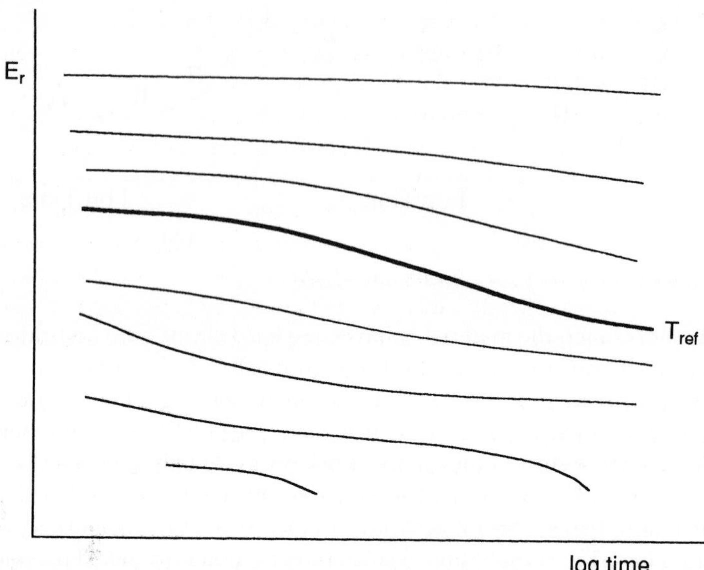

Figure 6.10 *Typical modulus–time relationships*

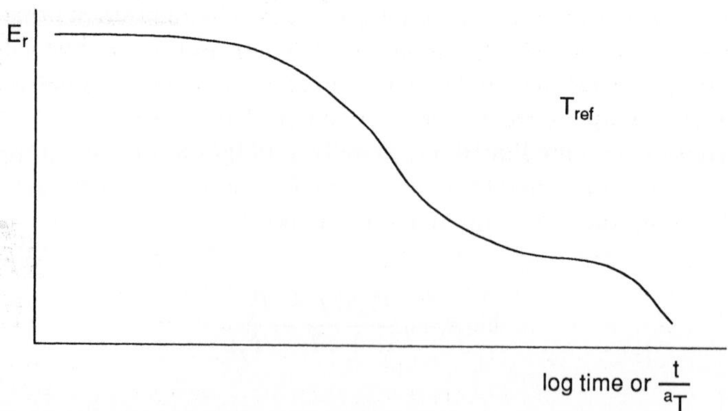

Figure 6.11 *Modulus–time/temperature master curve*

resistant coatings, the response to high speed impact can be gauged from the position on the master curve (relative to service temperature) corresponding to the frequency of the test impact, *i.e.* it is a hard brittle glass under these conditions, a viscoelastic solid, or a rubber.

Fracture Properties. For glassy polymers under tensile stress ultimate strengths and elongations are governed largely by the propagation of existing flaws and cracks. Many important classes of coatings are glassy materials and the majority are thermosets. Furthermore, close examination of the mechanical abuse scenarios reveal that the majority involve (at least in some part) the generation of tensile stresses in the coatings. Thus the fracture behaviour of thermoset glasses is an area of much effort and activity in coatings science. Whereas the science of fracture mechanics of polymers, in general, is well developed (see for example Kinloch and Young's text), it is less well understood for glassy thermosets. When these materials are used as thin films on relatively massive and stiff substrates little quantitative information is known.

Two material properties quantify the stability of a polymer against the initiation and propagation of cracks. These are the fracture energy, G_{1c} defined as the critical strain energy release rate (or the energy required to form unit area of crack) and fracture toughness, K_{1c} or critical stress intensity factor. K_{1c} relates the magnitude of stress intensity local to a crack in terms of loading and geometry. These properties can be measured for coatings systems by a number of techniques. For bulk specimens, the compact tension test geometry is favoured, for thin films single edge notched testing is often carried out. Fracture energy and fracture toughness are material constants and are influenced by temperature, rate of testing, and sample geometry. They are fundamental engineering properties of materials and their measurement for coatings allows comparison with published data for similar classes of polymer.

Published data also provide guidelines for the design of polymer and coating systems for toughness and formability. Toughness in glassy polymers is associated with the ability to undergo extensive yielding and plastic flow ahead of a flaw or crack tip. In bulk specimens, the material itself acts as a constraint to the formation of the plastic zone and relief at the edges of the sample is negligible. For thin films, edge effects are dominant allowing a larger plastic zone to develop. These two extremes of crack geometry are known as plane strain and plane stress conditions and account for the difference in toughness of glassy materials depending on sample size. Plane stress fracture toughness values are typically 2–3 times higher than those in plane strain. The problem remains as to how to treat thin coatings on large, stiff substrates since strategies for toughening of thermoset glasses are influenced by whether plane stress or plane strain conditions apply. At present neither the theory nor the practical techniques have been developed for determining the fracture behaviour of coated metals.

There are indications, however, that the body of information on the influence of polymer structure and properties on fracture behaviour can be related to the mechanical performance of coatings on their substrates. From fracture mechanics studies of thermosets, we would expect to see extensive plastic flow and deformation in the plastic zone ahead of the crack tip. Any structural features or conditions or properties which restrict this yielding and plastic flow will restrict the size of the zone (and hence the amount of energy absorbed there) and will render the material more susceptible to brittle failure. The following predictions can then be made which will lead to *decreased toughness*:

(i) increased modulus, E and yield stress, σ_y,
(ii) increased crosslink density, decreased $\bar{M}_c$,
(iii) decreased free volume, *e.g.* from physical ageing *q.v.*,
(iv) decreased temperature and increased rate of testing,
(v) increased network junction functionality,
(vi) increased proportion of network defects or elastically ineffective network chains and low extents of cure.

In terms of chain architecture the trends are less clear. It would seem that local main chain relaxation mechanisms [as seen in secondary or β transitions by Dynamic Mechanical Analysis (DMA)] are important in enhancing toughness. Detailed correlation, however, between polymer structure and yield, deformation, and fracture is not well developed.

Real coatings will often contain large amounts of pigment, fillers, and other additives which add further levels of complexity to the largely polymer based picture above. The addition of high modulus solids to polymers will increase the modulus and yield stress of the coating but this need not always serve to weaken it. Moderate levels of well dispersed fine pigment particles can act as reinforcement in terms of increased strength and improved toughness. Poor dispersion and wetting of pigments and fillers generally detracts from mechanical performance due to increased number of flaws and voids present. Similarly application and film formation processes which leave flaws in the coating will produce the same effect.

Internal Stresses. Application conditions and film formation processes can also lead to the establishment of significant internal stresses within coatings. The commonest causes are from solvent or water or other volatile release and from differential thermal expansion between coating and substrate. In the worst cases, the stresses are sufficient to cause cracking of the coating and loss of adhesion. Their presence however will inevitably detract from impact, formability, flexibility, and adhesion properties to some extent.

AGEING PROCESSES AND THE RETENTION OF PROPERTIES

The first part of this chapter dealt largely with the mechanical performance of coatings from the standpoint of design and development, to meet the stresses and strains imposed on them, in various end use applications. Whilst in many cases the critical mechanical abuse takes place soon after the coating is applied, there are many other examples where such abuse is experienced throughout the coating's service life or at some time afterwards. In such cases it is important to know the nature and the effect of the chemical and physical changes occurring in the coating as a result of exposure to its environment. It is then feasible to attempt to predict performance from suitably designed laboratory tests or from real time ageing or exposure experiments. As described in the first part of this chapter, it is necessary both to characterise the exposure environment accurately and to ensure that laboratory test protocol emulates that environment realistically. Similarly, it is necessary to be vigilant in checking the correlation between predictions from accelerated testing and real time experiments carried out by the coating laboratories, and also that these tests reflect performance in actual service.

This becomes increasingly important and more difficult as the design life of the coating lengthens. For example some coating systems for architectural steel cladding are required and specified to retain their appearance and corrosion protection performance for the design life of the building, say 15 to 25 years. Clearly the paint manufacturer cannot wait 25 years before offering a new product for this market. Furthermore, the cost implications of premature failure can be very significant thus the importance of a sound understanding and experimental approach is evident.

Exposure Environments

Weathering. Perhaps the most important category of exposure environment and its effect on paint performance is 'simple' outdoor weathering. It will be common to most people's experience that domestic outdoor paint work especially on wood does not retain its appearance or film integrity forever. The paint's gloss level decreases, colours fade or change, and cracking, peeling, and blistering may occur particularly in direct sunshine. On closer examination neither the exposure environment nor the chemical and physical processes going on are 'simple'. The paint is subjected to heat, oxygen, sunlight, water (liquid and vapour), atmospheric pollutants, and various cleaning chemicals. It will contain decreasing levels of residual solvent or water or both retained after application, as well as pigments, fillers, additives, or trace metal impurities incorporated during manufacture—all of which might have some photochemical activity. Furthermore, the paint itself will probably have experienced high temperatures during manufacture and, if it is a thermosetting chemistry, during curing or stoving (as used in prepainted metal frames, *etc.*) From this description, the thermal and photo-oxidative degradation processes of polymers used in coatings are going to be an important consideration in the design of exterior durable systems. The basic oxidation mechanism of organic polymers outlined in Scheme 1 has common features whether the source of initiation is thermal, photochemical, photophysical, or high energy radiation.

$$PH \rightarrow P^{\bullet} \qquad \text{Initiation}$$

$$\left.\begin{array}{lcl} P^{\bullet} + O_2 & \rightarrow & POO^{\bullet} \\ POO^{\bullet} + PH & \rightarrow & POOH + P^{\bullet} \\ POOH & \rightarrow & PO^{\bullet} + HO^{\bullet} \\ PO^{\bullet} + PH & \rightarrow & POH + P^{\bullet} \\ HO^{\bullet} + PH & \rightarrow & H_2O + P^{\bullet} \end{array}\right\} \text{Propagation}$$

$$\left.\begin{array}{lcl} 2P^{\bullet} & \rightarrow & P\text{—}P \\ 2POO^{\bullet} & \rightarrow & POOP + O_2 + \text{products} \\ P^{\bullet} + H & \rightarrow & PH \end{array}\right\} \text{Termination}$$

Scheme 1: Basic oxidation processes (P = Polymer)

An exact understanding of the sources of initiation in the oxidation scheme is not available. However, it seems likely that they would include one or more of the following; thermal processing and curing and/or intense mechanical shear giving rise to polymer radicals which go on to form groups or species which are themselves unstable to heat or sunlight. Chemical impurities introduced during manufacture and incorporated either into polymer or the formulation which are unstable to heat or sunlight. Whatever the origin, once initiated the oxidation proceeds largely through hydrogen abstraction in chain or branching reactions. Polymer structures which contain easily abstractable hydrogen will oxidise more rapidly and so these should be avoided for exterior durable systems. For example, tertiary and allylic hydrogens are particularly susceptible to abstraction by free radicals (see Scheme 2—note the similarity to the oxidative curing mechanism of alkyds discussed in Chapter 7). Similarly structures which strongly absorb the short wavelength, high energy portion of solar radiation should be avoided. In such cases the energy can be sufficient to cause photochemical reactions leading to chain scission, radical formation, and the formation of other chromophores. Epoxy resins based on bisphenol A, for example, deteriorate fairly rapidly on exposure to sunlight due to chain scission and phenoxy radical formation.

$$\mathrm{R{-}\underset{\displaystyle R}{\overset{\displaystyle H}{C}}{-}R} \xrightarrow{I^{\bullet}} \mathrm{R{-}\underset{\displaystyle R}{\dot{C}}{-}R}$$

$$\mathrm{{-}\overset{\displaystyle H}{C}H{-}CH{=}CH{-}} \xrightarrow{I^{\bullet}} \mathrm{{-}\dot{C}H{-}CH{=}CH{-}}$$

Scheme 2: Easy abstraction of $H^{\bullet}$ from methine or allylic groups

The overall effect of the complex reactions will be predominantly chain scission or crosslinking depending on the chemical structure. Chain scission results in a gradual softening of the coating with a lowering of T_g and as MW decreases embrittlement and eventual solubilisation of thermosets will take place. Where crosslinking reactions dominate then a stiffening and increasing T_g is observed often with a consequent embrittlement (although there may be an initial strengthening of the material on crosslinking, excessive crosslinking inevitably leads to embrittlement). In either case, the extent of damage to the polymer can be such that material at the surface can be lost through solubilisation or attrition due to inability to sustain even the low strains associated with differential thermal expansion. The net result will be gradual erosion and roughening of polymer from the surface with consequent loss of gloss. Pigment or filler particles may be exposed as the binder is lost and these particles can be easily removed from the surface—the phenomenon known as chalking.

The pigment particles may also be affected by solar radiation, especially the organic types. When mixtures of these pigments are used to produce a specific colour and one of these is more susceptible to radiation, a gradual colour change can take place on exposure as one of the pigments undergoes photochemical reactions. There are some pigments which are themselves photochemically active. A well known example is titanium dioxide which in the presence of sunlight, oxygen, and water will generate hydrogen peroxide and radical species. To reduce this activity, specific forms of titania with various surface treatments are employed and as a result exterior durable white paints based on titania pigmentation are widely used.

For some systems, exposure to high humidity or water immersion can cause serious deterioration in the properties of coatings either through chemical attack (hydrolysis) or through the dilational strains caused by water uptake. In many cases the combined effect of water and sunlight represents a more severe environment than either separately. For this reason a number of laboratory accelerated test regimes have been designed to produce exposure of coatings to various cycles of temperature irradiation and humidity. Similarly, a number of test sites, notably Florida, are widely used by the coatings industries and specifiers of coatings as severe natural exposure environments. Coating systems, for exterior applications, are commonly quoted as being 1 year or 5 years Florida which is shorthand for the retention of 50% of original gloss after those periods of exposure. There still remains the problem of the meaning of 5 years Florida data for coatings which will be used in Northern Europe (say). Almost certainly they will retain their properties for longer than 5 years, but for how much longer?

There have been many attempts to produce an accelerated laboratory weathering test for predicting outdoor durability. As described above many incorporate cycles of dry heat, humid heating, irradiated, and dark periods of varying length. It is perhaps unfair to try to summarise the vast amount of effort put into accelerated weathering tests. In general however the correlation between these accelerated tests and real time exposure is not very good. To be of any use the test needs to give a result in an acceptably short experimental time say 1–2 months. In order to produce an acceleration factor of approximately 50 (in terms of reproducing 5 years Florida data in 5 weeks) then it is necessary to increase temperature, relative humidity, and the intensity of incident radiation. For some time it was common to use radiation of shorter wavelength than occurred naturally in sunlight. This certainly produced rapid results, but they were unrelated to what happened on natural exposure. The emphasis now is on the use of filtered UV sources which cut off at 300 nm (as does natural sunlight), but which deliver higher intensity illumination than sunlight at the shorter wavelengths. Similarly, too high a test temperature may produce erroneous results if, for example, the coatings T_g is exceeded in the test but not in actual service. It has also been found that dark 'rest' periods are necessary to enable oxygen concentrations in the surface layer to be replenished as happens naturally at night time. Thus, continuous exposure to light and oxygen may in fact run in oxygen starved conditions and hence give misleading durability predictions.

Accelerated weathering tests are still very widely used in the coatings industries despite the observations made above and the realisation that they only provide an indication of the performance rather than any predictive capability. Increasing effort has been expended on developing methods of increased analytical sensitivity so that the slight changes occurring after relatively short natural exposures can be monitored accurately. This approach will give quantitative data on some aspect of the weathering process to allow kinetic analysis and the development of models to predict long term performance. This has been successfully carried out by Ford for automotive paints using electron spin resonance (ESR) spectroscopy to follow free radical processes and also using titrimetric determination of hydroperoxide, an important species in the oxidation scheme. Other techniques examined in this connection are FTIR and chemiluminescence measurements.

Stabilisation. There are relatively few examples where stabilisation by additives has been used successfully in coatings. For pigmented systems, stabilisers against thermal- and photo-oxidation need to be effective in the surface resin rich layer which determines gloss levels. Once this layer has disappeared or been damaged then appearance properties are lost and so any stabilisers need be present only in the surface. These additives will have a finite solubility in the resin and, since they may be consumed in time, protection of the paint will be lost at some stage. Ideally replenishment of the additive via migration from the bulk to the surface at the right rate would provide long term protection. This has proved difficult to achieve in practice and the use of UV stabilisers, antioxidants, *etc.* in coatings is not as widespread as it is in the plastics and rubber industries.

One example where stabilisation has proved successful is in automotive paints where a clear coating is often used over a pigmented base coat for aesthetic reasons. In the early stages of development, the clear topcoat was chosen for its very low UV absorption characteristics and long term durability. Less attention was paid to the chemistry and durability of the basecoat and in service loss of appearance through disruption of the topcoat–basecoat interface was observed. In this case the topcoat was essentially transparent to UV which then caused degradation of the basecoat. The use of UV absorbers and stabilising additives in the topcoat successfully prevented UV penetration to the basecoat and achieved long term exterior durability for the system as a whole.

Physical Ageing

Physical ageing is the name given to the slow changes which occur in glasses with time. In contrast to the chemical ageing processes described above, these changes are reversible when the glass is heated to above its T_g. The origin of this ageing phenomenon lies in the non-equilibrium nature of glasses. When amorphous polymers are cooled through their T_g, they solidify and stiffen to form glasses. In this glassy state, rates of structural relaxation become very slow and polymer conformations are fixed in a non-equilibrium state. If the temperature is held at some value below T_g, the process of relaxation towards equilibrium configuration continues and as a result many material properties change with time. The effects are most pronounced on low strain and low frequency properties such as creep and stress relaxation rates which decrease on ageing. There are, however, significant effects on other important properties notably elastic moduli, yield stress, and density all increase on ageing and as a result impact resistance, ductility, and formability all decrease.

The process of physical ageing is highly temperature dependent, being fastest at temperatures closest to T_g. Similarly, the rate of cooling through T_g is important in terms of the amount of excess free volume trapped in the glass with rapid or quench cooling leaving the material furthest from its equilibrium state. In terms of a coating's mechanical performance and the measurement of material properties, it is clearly important to know and understand the effect of thermal history, especially in terms of designing test protocols for coatings which are stoved above T_g and those which have a T_g within *ca.* 30 °C of ambient.

The phenomenon is well illustrated by the following example in which powder coating of metal pipes of different wall thicknesses gave different performance properties. Here a single powder paint was used to coat a variety of steel pipes under similar conditions of film thickness, stoving times and temperatures, and water quenching. Samples were cut from the pipes soon after coating and bent to assess flexibility of the coatings. Those from thin walled pipes passed the bend test whereas those from thicker walled pipe failed. When tested the next day, samples from both types of pipe failed the test. The reason for the differences in performance arose from the cooling rates of the two types of pipe in the water quench. The high mass of metal in the thicker walled pipe led to a slow rate of cooling and hence more physical ageing took place on cooling to ambient. The thin walled pipe coating was essentially quenched under the same conditions and so was in its toughest, most formable condition when tested after coating. Storage overnight led to a similar amount of ageing to that experienced by the coating on thick walled pipe and this produced failure on bending. The lessons here are that proper sample conditioning and control of thermal history are essential if performance is to be reflected accurately in laboratory tests for glassy materials.

CHEMICAL EXPOSURE

Some of the most technologically demanding applications for coatings occur in the protection of structures against chemical attack. The range of end uses where this type of performance is required is wide. In terms of size, the structures vary from a food or drinks can holding less than half a litre to a ship's chemical tank with a capacity of a million litres or more. Similarly, the chemical exposure can be in the form of gas, liquid, or solid, organic or inorganic, hot or cold, acidic, neutral, or basic. The exposure may be continuous or intermittent and in some cases such as chemical cargo tanks there may be a succession of different chemicals.

In general, we can separate the chemical resistance performance of a coating into two broad categories. In the first, chemical resistance is required where coatings suffer occasional exposure to some form of chemical stress. They must nevertheless continue to perform their other functions during and after this exposure. Three examples serve to illustrate this type of chemical attack and are given in order of increasing severity. Coatings for washing machines need to resist hot or boiling water and solutions of soaps and detergents. Car paints also need to cope with soap and water but resistance to hydrocarbons (as fuels, greases, and lubricants) and hydraulic fluids will be required. Finally, aviation finishes need, in addition to many other exacting performance requirements, to be resistant to prolonged exposure to aircraft hydraulic fluids and lubricants.

In the second broad category, the primary function of the coating is to provide corrosion protection for the substrate or structure against long term exposure to a chemically aggressive environment. The examples of food containers and chemical tankers have already been described, but more common is the protection of steel from simple atmospheric or saltwater attack.

For either category the exposure to some form of chemical environment must not cause the coating to be lost through dissolution or disbondment or by becoming mechanically weak from absorption of fluids. Where prolonged chemical exposure is expected care in the selection of coating chemistry must be exercised to avoid the possibility of direct chemical attack or reaction. Polyesters would not normally be used, for example, as coatings for structures exposed to strong aqueous bases or acids. Here hydrolysis of main chain ester groups can lead to breakdown and eventual dissolution of the coating. Similarly, hydrocarbon coatings will be attacked by strong oxidising species.

Amorphous thermoplastics will generally be soluble in a particular range of organic solvents dependent on solubility parameters. For this reason thermoset or crosslinked coatings are widely used where resistance to a broad spectrum of chemicals is required. In the absence of direct chemical attack or specific reactions, crosslinked coatings will be insoluble in all solvents. They will absorb solvents, however, which leads to swelling, softening, and weakening. The amount by which crosslinked polymers swell depends on a number of factors of which the two most important are crosslink density and the level of interaction between polymer and solvent. Highly crosslinked structures take up less solvent than lightly crosslinked ones and this forms the basis of the familiar solvent swelling technique used to characterise networks. In general, good solvents for the polymers before crosslinking will be the most powerful swellers for the final network and hence taken up in the largest amounts.

The consequence of solvent uptake is that stress is exerted on the adhesive interface between a coating and its substrate through the dilational swelling strains. Disruption of the interface may also occur if solvent–substrate interaction is favoured over that of the polymer. Whilst this may be useful for solvent based paint strippers in the home, it is clearly going to be a severe nuisance for a coating in a chemical environment. Solvent uptake will also lower a coating's T_g and, in this softened and swollen condition, tensile and tear strengths will be much reduced and so its resistance to mechanical damage will be impaired. When a coating on a land or ship based storage tank is exposed to a succession of different contents, solvent retained from one may be released into a second. This presents obvious problems when such contamination is unacceptable, for example when foodstuffs are carried after chemical cargoes in a ship's tank. In such cases the sequence of cargoes is strictly controlled.

There have been three traditional approaches to providing resistance to chemical uptake and swelling. Where a single or dominant type of chemical exposure is expected, coatings architectures have been chosen which are markedly different from the chemical, *i.e.* polar polymers for non-polar solvents and *vice versa*. The second strategy is to obtain the highest possible crosslink density (commensurate with other required properties) in the coating for applications where a broad spectrum of chemical resistance is required. For reasons outlined in Chapter 5, high extents of cure may be difficult to obtain in such systems especially when only ambient curing conditions are possible. In consequence, there may be a significant sol fraction remaining in the coating which will be extractable in some solvents or chemicals. The last approach has been to reduce the concentration of swellable polymer in the coating by the use of fillers and pigments. This has implications for the viscosity of the coating at application and is becoming less attractive with the drive to reduce solvent levels in coatings.

The subject of corrosion protection by coatings remains somewhat controversial despite its obvious importance and the large and continuing research effort devoted to its understanding. In its simplest aspect, coatings provide protection for metal against corrosion by acting as a barrier to the reactants or products of corrosion reactions. Typically these species would be oxygen, water, hydrogen, ions, or electrons.

One body of opinion maintains that organic coatings do not provide a sufficiently impermeable barrier to oxygen or water to prevent corrosion at the metal substrate. The rates of diffusion of these species in perfect, flaw free organic films would be sufficient to support the observed rates of corrosion of unprotected metals. It further maintains that the fact that coatings can provide very effective corrosion protection is due to the very low permeability of ionic species or electrons in the covalent, organic, low dielectric medium of the coating. To account for the corrosion protection performance of real coatings the presence of flaws, imperfections, and structural heterogeneities of various sorts is invoked. These flaws can be macro- or microscopic or at the macromolecular level and are held to be responsible for providing pathways for the species above to sustain local corrosion centres.

In support of the above proposition it is clear that the quality of coating application and film formation is of major importance in corrosion performance. Where this quality is poor and flaws, voids, and dewetted areas are detectable by eye or microscope, then corrosion performance is impaired. Multiple paint applications will often improve performance when the chance of coincident flaws is much reduced. Similarly, the addition of non-polar, low dielectric additives such as hydrocarbon waxes and tars has been used to improve corrosion resistance. It should also be said that oxygen/water barrier enhancing fillers such as glass flake, mica, micaceous iron oxide have also been effective in anticorrosive coatings. However in the absence of microscopic flaws, *etc.* the evidence for heterogeneities at the molecular or supramolecular level being responsible for corrosion pathways is much less convincing—although still a popular opening supposition by corrosion scientists.

SUMMARY

The emphasis of this chapter has been to describe the links between coating performance in service and the tests and procedures which are used to develop materials to meet a required performance. Attempts to extend these relationships through to polymer structure have also been described. It is clearly an ambitious and daunting task to design coatings for specific end uses starting from a polymer architecture. In practice, of course, it is usual for a succession of people or teams to take a project from materials to commercial products. The key to success is the effective generation and transfer of quality information on what the coatings need to do, what they actually do, what makes them perform the way they do, and hence what can be done to change performance.

BIBLIOGRAPHY

L. E. Nielsen, 'Mechanical Properties of Polymers and Composites', Dekker, New York, 1974.

A.J. Kinloch and R.J. Young, 'Fracture Behaviour of Polymers', Elsevier, London, 1983.

Chapter 7

Coatings Polymers—Thermoplastic and Reactive Systems

A. R. MARRION

POLYMERIC MATERIALS

Polymers are large molecules, usually constructed of one or more repeating units or 'mers' derived from monomers. Their chemistry is extremely diverse (as we shall see) but their character is also heavily dependent on their molecular weight and architecture.

Representing a monomer as M, its polymers can be straight chains (1), branched (2), radiate (3) (stars if the number of branches exceeds about four), or even cyclic (4). When a second monomer M′ is introduced it is possible to envisage random (5), alternating (6), blocky (7), or grafted (8) copolymers.

The molecular weights of polymers can vary from a few thousand to several million and it is often more convenient to think in terms of degree of polymerisation (d.p.), *i.e.* the number of monomer residues.

No practical polymer system is composed entirely of molecules of identical molecular weight; the molecular weight is an average and can be expressed in a number of ways, *e.g.* the number average

$$M_n = \frac{\Sigma n_i m_i}{\Sigma n_i} \tag{1}$$

or the weight average

$$M_w = \frac{\Sigma w_i m_i}{\Sigma w_i} = \frac{\Sigma n_i m_i^2}{\Sigma n_i m_i} \tag{2}$$

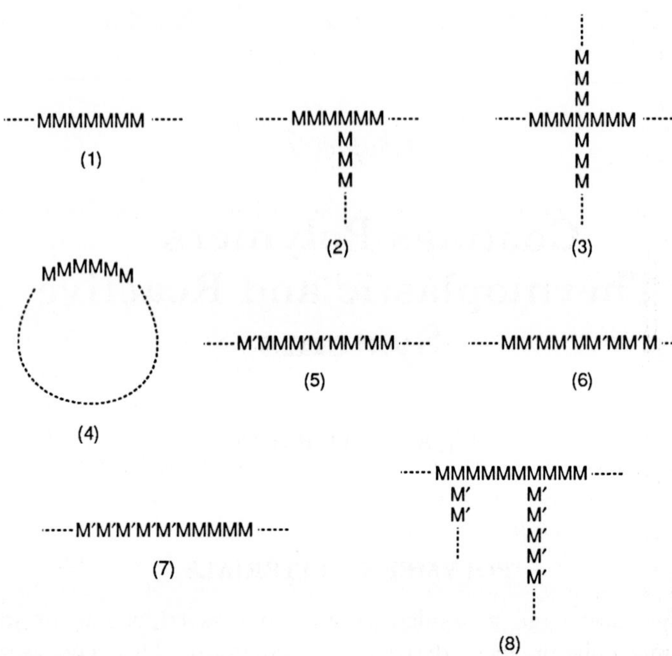

(where n_i is the number of molecules of mass m_i and w_i is the weight of such molecules). The ratio $\frac{M_w}{M_n}$ is defined as the polydispersity, d, of a particular polymer and is also an important characteristic.

The manufacture of polymers involves either a chain growth mechanism (Scheme 1), in which monomers add to a growing chain.

$$\mathrm{M + M \longrightarrow MM \xrightarrow{M} MMM \xrightarrow{M} MMMM \xrightarrow{M}} \textit{etc.}$$

Scheme 1

or a 'step growth' mechanism (Scheme 2), in which monomers link together in small groups which in turn link to each other producing an exponential increase in molecular weight.

$$\begin{aligned} \mathrm{M + M} &\longrightarrow \mathrm{MM} \\ \mathrm{MM + M} &\longrightarrow \mathrm{MMM} \\ \mathrm{MM + MM} &\longrightarrow \mathrm{MMMM} \\ \mathrm{MMMM + MMM} &\longrightarrow \mathrm{MMMMMMM} \\ &\textit{etc.} \end{aligned}$$

Scheme 2

The polyesterification reaction between a diol and a dicarboxylic acid (Scheme 3), is representative of the step growth process.

$$n\,\text{HOOC}\sim\sim\text{COOH} + (n+1)\text{HO}\sim\sim\text{OH} \xrightarrow{-2n\,H_2O} \text{HO}\sim\sim\text{O}\left[\text{OC}\sim\sim\text{COO}\sim\sim\text{O}\right]_n\text{H}$$

Scheme 3

The product necessarily contains alternating diacid and diol residues and its end-functionality and molecular weight are approximately determined by the starting stoicheiometry.

At the outset, there is a plentiful supply of reactive groups, and esterification occurs readily. As the reaction proceeds, molecular weight increases and reactive group concentration becomes lower. In practice it is difficult to drive any step growth process to completion, and high temperatures (250 °C), catalyst, and continuous removal of the condensed species may be needed to achieve a satisfactory conversion.

Combinations of different acid and polyol monomers are frequently used, for example the 'hard' phthalic acid might be combined with the 'soft' adipic acid and a suitable polyol to obtain a satisfactory glass transition temperature (T_g).

A degree of branching can be introduced by using low levels of tri- or higher functional acid or alcohol though the stoicheiometry then becomes much more complex and the polydispersity is inclined to be high. Formulations are best derived by computerised iteration, which can also give warning of potentially gellable compositions.

Some of the more usual polyester monomers are listed in Table 7.1.

Other step growth processes do not necessarily involve condensation chemistry.

Polymerisation of 'acrylate' monomers, under free radical conditions, is an example of a chain growth process (Scheme 4).

$$\text{I}^\bullet + \text{CH}_2{=}\text{CH–COOR} \longrightarrow \text{I–CH}_2\text{–}\dot{\text{C}}\text{H–COOR} \xrightarrow{n\ \text{CH}_2=\text{CH–COOR}} \text{I–CH}_2\text{–CH(COOR)–}[\text{CH}_2\text{–CH(COOR)}]_{n-1}\text{–CH}_2\text{–}\dot{\text{C}}\text{H(COOR)}$$

Scheme 4

Acrylate monomers are olefinic compounds of enhanced reactivity including acrylic acid derivatives and related compounds and a few 'honorary' acrylics such as styrene (*c.f.* Table 7.2).

The initiator, I$^\bullet$, is a free radical derived by homolysis of a species such as a peroxide or an azo compound, or by a redox reaction (Scheme 5).

$$H_2O_2 + Fe^{2+} \rightarrow {}^\bullet OH + {}^-OH + Fe^{3+}$$

Scheme 5

Table 7.1 *Typical polyester monomers*

Group	*Structure*	*Name*
'Soft' diacids	$HOOC-(CH_2)_n-COOH$	Alkanedioic Acids (n = 0–10)
	e.g. $HOOC-(CH_2)_4-COOH$	Adipic acid
	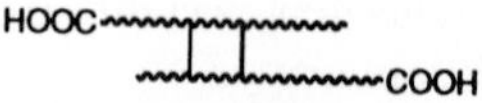 HOOC … COOH	Dimerised fatty acid (n = *ca.* 34)
'Hard' diacids	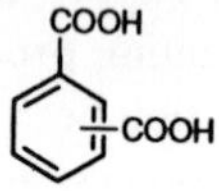 COOH, COOH	1,2- = *o*-phthalic acid (anhydride)
		1,3- = isophthalic acid
		1,4- = terephthalic acid
	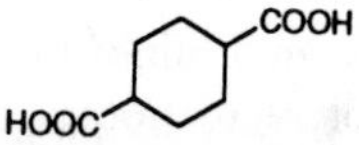 HOOC, COOH	Cyclohexanedicarboxylic acid
Triacids	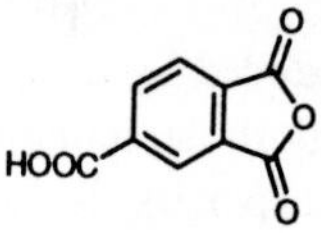 HOOC, O, O, O	Trimellitic anhydride
Heterofunctional acids	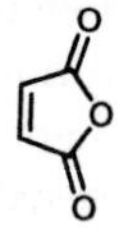 O, O, O	Maleic anhydride
Diols	$HO-(CH_2)_n-OH$	Alkanediols (n = 2–10)
	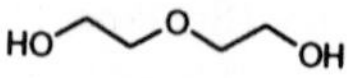 HO, O, OH	Diethyleneglycol
	 HO, OH	Neopentylglycol
Triols	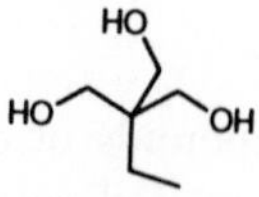 HO, HO, OH	Trimethylolpropane
	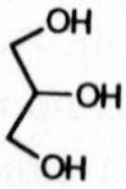 OH, OH, OH	Glycerol
Tetrols	 HO, OH, HO, OH	Pentaerythritol

Table 7.2 *Some typical 'acrylic' monomers*

Structure	*R*	*R′*	*Name*
R-substituted vinyl, COOR′	H	H	Acrylic acid
	Methyl	H	Methacrylic acid
	H	Methyl Ethyl Butyl	Alkyl acrylate
	Methyl	Methyl Ethyl Butyl	Alkyl methacrylate
R-substituted vinyl, $CONH_2$	H	—	Acrylamide
	Methyl	—	Methacrylamide
R-substituted vinyl, C≡N	H	—	Acrylonitrile
	Methyl	—	Methylacrylonitrile
(styrene structure)	—	—	Styrene

Chains grow rapidly to a molecular weight determined by the ratio of monomer to initiator, unless opportunities for chain transfer exist. Chain transfer occurs when the growing macroradical abstracts a moiety (usually $H^{\bullet}$) from a species in the reaction mixture (Scheme 6). A new radical is left behind to initiate a new chain (Scheme 7). Solvent, monomer, or polymer can act as chain transfer agents but it is common practice to add mercaptans to reduce molecular weight and economise on initiator.

$$I\sim\sim\sim^{\bullet} + HSR \longrightarrow I\sim\sim\sim H + RS^{\bullet}$$

Scheme 6

$$RS^{\bullet} + n\ CH_2{=}CH\text{–}COOR \longrightarrow RS\sim\sim\sim^{\bullet} \longrightarrow \textit{etc.}$$

Scheme 7

Reaction chains can also be terminated by the combination of two macroradicals or much less frequently, by disproportionation, when one macroradical acquires $H^{\bullet}$ from another, leaving an unsaturated end group.

Products of free-radical polymerisation of different acrylic monomers are more or less blocky or random in residue distribution, according to the reactivity of each monomer, with macroradicals tipped with like or unlike residues ('reactivity ratios').

Chain growth processes can also be initiated by cationic or anionic species. In some cases chains with 'living' end groups can be produced, and will initiate polymerisation of a different monomer so that well 'tailored' block copolymers can be produced.

The geminal substitution on a polymethacrylate produces a much stiffer backbone than the equivalent polyacrylate, hence methacrylate monomers confer high T_gs on their polymers. A coatings formulation may contain a combination of methacrylates and acrylates to achieve the desired hardness, with small amounts of acid or amide to improve adhesion. A wide range of other 'functional' monomers can also be employed to provide crosslinking sites (*c.f.* Table 7.6).

POLYMERS IN COATINGS

Organic coatings are formulated from a number of ingredients, usually including pigments and solvents, but the one indispensable component is a polymer or polymer precursor to act as a 'binder' or 'vehicle'. The characteristics of the polymeric binder usually determine the character of the coating, so that polyurethane paints are noted for their toughness, epoxies for their adhesion, and so on. They also have a major bearing on the method of application and conditions of film formation.

Many coatings polymers are manufactured 'in house' by the coatings manufacturers, according to their own technology. Others are prepared by specialist suppliers, either because they have wide application and economics of scale can be exploited, or because their manufacture requires particular plant, or control, or is proprietary.

A coating binder in service is required to be indifferent to varying extents, to mechanical abuse, chemical insults, and the effects of its environment, yet be applied easily under mild conditions. It is necessary to convert a liquid or fusible solid into an intractable solid on demand, and that transformation is a key feature of any successful coatings system. Several of the transformation strategies in widespread use are summarised in Table 7.3.

Dissolution of a relatively high molecular weight polymer in a suitable solvent, whose loss by evaporation provides a thermoplastic coating, can give a satisfactory lacquer. However, a polymer of sufficient molecular weight tends to provide viscous solutions with the result that high dilution is required.

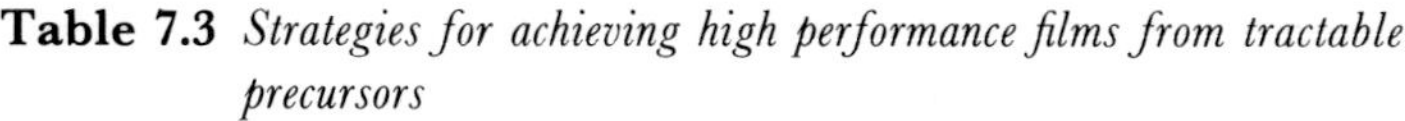

Table 7.3 *Strategies for achieving high performance films from tractable precursors*

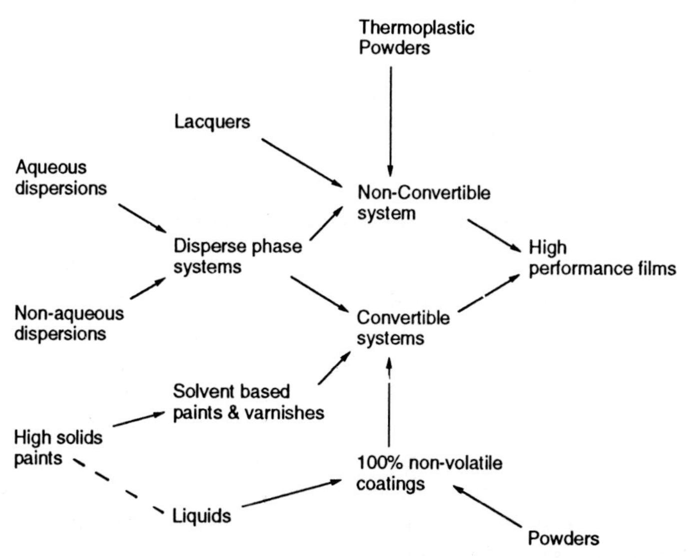

The economic disadvantages of lacquers have long been recognised; not only is a large proportion of the composition ultimately lost to the atmosphere, but the film left behind is likely to be much thinner than desired and multiple applications may be necessary. In a similar way, thermoplastic powders are generally too viscous in the melt to coalesce to films of satisfactory integrity and appearance in other than very thick layers.

Reactive systems can be based on lower molecular weight precursors which 'crosslink' or 'cure' after application. Such systems are described as 'convertible' or 'thermosetting'. They can be designed with substantially increased non-volatile contents and usually show enhanced mechanical and chemical properties as compared with non-convertible systems.

More recently, awareness of the environmental damage which emission of some organic solvents can inflict, has stimulated energetic development of systems requiring less, or none (*c.f.* Chapter 2). High solids systems are extending the reactive polymer concept further each year with lower molecular weight oligomers, intrinsically more fluid systems and, in some cases, reactive solvents. The ultimate conclusion—the solvent-free system has been achieved but such systems, and to a lesser extent high solids coatings, present severe challenges in application (*c.f.* Chapter 4).

When high stoving temperatures can be used, thermosetting powder coatings provide a useful solution since the solid composition can be handled and sprayed as a fine powder.

An alternative approach is to provide the polymer as a solution or dispersion in an unexceptionable liquid such as water, selected organic solvents, or even supercritical carbon dioxide. It will be dealt with in Chapter 8.

THERMOPLASTIC BINDERS

The fact that film formation by lacquer drying is a reversible process, and the polymer can be removed with the same solvent is advantageous when temporary coverage is required, as when coating finger nails, and when cleaning or repair of localised damage is called for. It also means that intercoat adhesion can be very great since some re-dissolution on overcoating can occur, but the tendency of partially crosslinked systems to swell and disbond is absent.

However, a molecular weight in the order of 10^5 is needed to achieve acceptable performance, and such materials are often diluted to 10–15% non-volatiles with the problems noted above.

Cellulose Derivatives

Derivatives of cellulose are amongst the most important thermoplastic coatings polymers.

Cellulose itself occurs widely in nature. It is a polymer of anhydroglucose units(9) and exists as crystalline, hydrogen bonded fibres. Esterification of some of the three hydroxyl groups per residue leads to organic solvent soluble polymers.

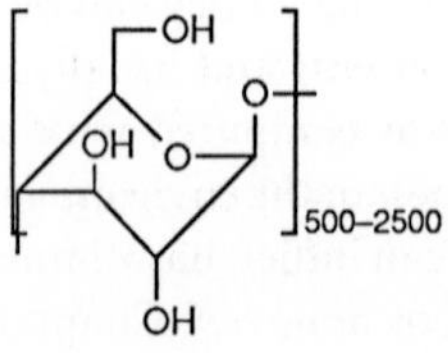

(9)

'Nitrocellulose' (*i.e.* cellulose nitrate) is prepared by treating cotton wool, wood pulp, or other source of cellulose with nitric and sulfuric acids. Levels of nitration varying between 2 and 3 units per residue are attainable and the products are dissolved in strong solvents such as esters. They were originally used in aircraft 'dope' to stiffen and tighten the canvas covering, and in automotive finishes where their relative softness allowed them to be polished to a high gloss. In today's rather changed world they are used in car refinish, furniture finishes, and inks, usually in conjunction with plasticisers and other polymers to provide the required mechanical properties and increased non-volatile content. Though their use has declined considerably they still occupy a significant position.

The mixed cellulose ester, cellulose acetate/butyrate, in several grades according to composition, has widespread importance. It shows unusual compatibility with other polymers and exerts a profound effect on their rheology. In coatings containing metal flake it promotes flake alignment and is frequently added in quite high proportions to automotive 'base-clear' base coats.

Other cellulose esters and ethers have limited value as thickeners in water-based compositions, for example.

Acrylics and Vinyls

Thermoplastic acrylics, usually predominantly poly(methyl methacrylate) of molecular weight in the order of 100 000 have been used as automotive topcoats of the flow and reflow type. They are sanded to remove imperfections after stoving at a low temperature, then re-stoved at a higher temperature to restore a high gloss.

Alternatively, they can film-form at ambient temperature when sufficiently plasticised either 'internally' with co-monomers such as butyl acrylate or with an external additive such as an ester or certain proprietary chlorinated materials. Such compositions are used as marine top coats where their retention of the usual white colour and gloss is invaluable. Polystyrene has found limited acceptance in zinc-rich anti-corrosive primers.

The best known vinyl polymers are prepared by free radical copolymerisation of vinyl chloride (10) and vinyl acetate (11) to achieve the required hardness (homopolymers of (10) being too hard, whilst those of (11) are too soft).

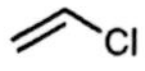

Cl (10)

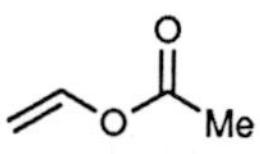

O Me O (11)

Small amounts of functional monomers such as maleic anhydride are frequently added to promote adhesion, or introduce a low level of crosslinking.

Vinyl polymers are typically highly water resistant and have been used alone as can-linings or combined with other thermoplastic or thermosetting polymers such as acrylics, epoxies, or hydrocarbon polymers for marine paints.

Hydrocarbon or petroleum polymers are low cost materials made by cationic or free-radical polymerisation of unsaturated materials such a coumarone (12) and indene (13) which are by-products of coal tar or petroleum refining. They are analogous to polystyrene in much of their behaviour.

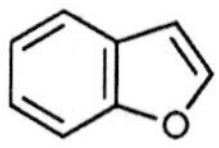

(12) 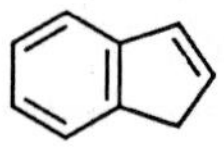 (13)

Pitch at levels to about 50% confers excellent properties on vinyl or epoxy compositions where its black colour can be tolerated, but unfortunately tends to contain mutagenic condensed aromatic hydrocarbons and has lost favour.

Poly(vinyl alcohol) is prepared by hydrolysis of polyvinyl acetate, since vinyl alcohol exists predominantly as acetaldehyde. The hydrolysis is never complete and the properties of different grades are dependent on levels of residual acetyl groups. It and polyvinyl ethers, such as poly(vinylisobutyl ether), are used as alternative plasticising co-monomers with vinyl chloride.

Polyvinylbutyral is made by treating poly(vinyl alcohol) with butyraldehyde and contains the structure (14) and residual hydroxyl groups, as well as the acetate groups in the original polyvinyl alcohol.

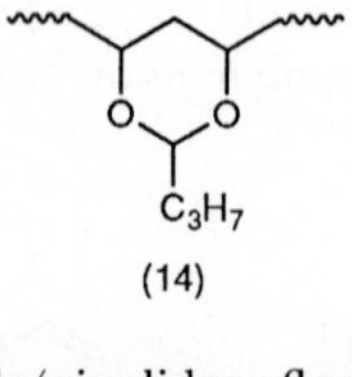

O O

C_3H_7

(14)

Polyvinyl fluoride and poly(vinylidene fluoride) are high cost materials with outstanding weatherability and water resistance. They are increasingly used in coil coating for exterior use.

Chlorinated Polymers

There is a wide range of chlorinated polymers available and in use, besides polyvinyl chloride, though chlorinated polymers are attracting a certain amount of disfavour on the grounds that incineration of waste material or material involved in fires may give rise to toxic fumes.

Chlorinated rubbers are highly chemically resistant media though liable to decompose at only moderately elevated temperatures. They are made by treating rubber with chlorine when the expected addition and substitution, and more complex reactions besides, take place with progressive hardening. The most heavily chlorinated systems contain 65% of chlorine. Chlorinated rubber media are used, for example, in high build primers for ships.

Analogous chlorinated systems are chlorinated polyethylene, chlorinated polypropylene, chlorinated ethylene/vinyl acetate copolymers, and even chlorinated polyvinyl chloride.

REACTIVE BINDERS

Crosslinking Strategies

The architectural strategies available for crosslinking polymers are summarised in Table 7.4. Where A and B represent reactive functional groups and each molecule is invested with at least two functional groups, and some have three, complete reaction leads to an infinite network irrespective of the original size of the reacting molecules. It is conventional to regard a polymer with reactive groups as a 'reactive binder' and a small molecule as a 'curing agent'.

Table 7.4 *Crosslinking strategies using co-reactive 'A' and 'B' functions*

A-functions	*B-functions*
Film forming polymer	Low molecular weight molecule ('Curing agent')
Film forming polymer	Film forming polymer
Single film-forming polymer	
Low molecular weight molecule	Low molecular weight molecule

There is no clear-cut distinction between 'reactive binders' and 'crosslinking agents' or indeed some polymer types and their crosslinking chemistry. To avoid undue repetition this section will confide itself to the general chemistry of reactive polyesters and acrylics whilst certain very important polymers such as alkyds, epoxies, and melamine formaldehydes will be discussed under the appropriate crosslinking chemistry.

The reaction of a functional polymer with a low molecular weight curing agent is frequently used. The advantage of such systems is that the film former can be prepared economically with the most convenient functional groups, and any 'difficult' chemistry confined to a relatively low volume material, free from the constraints of polymer preparation. The precise functionality of one component at least is then known.

The interaction of two reactive polymers is used to achieve networks with character derived from the different polymers, for example greatly improved mechanical properties can be obtained by crosslinking a stiff laterally functional polymer with a flexible telechelic crosslinker. It is sometimes also necessary to polymerise a low molecular weight crosslinker, or form its adduct with a polymer to reduce its volatility and hence toxic hazard. In practice, many systems use oligomeric crosslinkers of intermediate type.

The use of ambifunctional polymers is advocated from time to time as a way of fixing crosslinking stoicheiometry from the outset, but is not generally favoured since it allows maximum opportunity for premature reaction, and denies the formulator the opportunity of varying the composition. Notable exceptions, however, are oxidising alkyds where A and B may be identical or similar free radicals.

Film formation from two low molecular weight precursors has the theoretical attraction of leading to well-defined networks, and offering high solids but is fraught with difficulties. In most practical cases at least one component is oligomeric.

Polymer Synthesis and Functional Group Distribution

The crosslinking characteristics of our two model coatings polymers are compared in Table 7.5.

Step growth polymers are terminated with residues of the functional

Table 7.5 *Typical characteristics of step growth and chain growth polymers*

	Step growth (Polyester)	*Chain growth (Polyacrylate)*
Molecular weight	'Low' easily	'High' easily
Functionality	< about 4	> about 4
Functional architecture	Terminal	Lateral
Functional chemistry	That of excess reagent, *e.g.* —OH, —COOH	Variable almost at will

groups involved in their synthesis unless deliberately end-capped. For example, a linear polyester prepared with excess diol would be expected to be hydroxyl tipped. Unfortunately, complete reaction is never attained and such polyesters will also contain small quantities of acid end group; it will not be satisfactorily 'telechelic', and its network properties when crosslinked will be deficient. The usual remedy is to introduce a low level of branching as noted above. The notional functionality is then somewhat higher than 2. Attempts to raise the functionality much above 3 usually result in gellation.

It is equally possible to prepare acid tipped polyesters by using an excess of diacid, or to tip a preformed polyester with another functional group such as isocyanate. It is possible with care to react a difunctional moiety so as to produce tips with minimal chain extension (*e.g.* Scheme 8). The polyurethanes employed in the coatings industry are such materials containing flexible polyester or polyether blocks since a pure step growth polyurethane would tend to be intractable.

~~OH + NCO NCO → ~~O NH NCO Me O

Scheme 8

Many of the other reactive polymers used in the coatings industry are of a step growth type, but are more conveniently considered under their respective cure chemistries.

Chain growth polyacrylates prepared by free-radical processes are usually laterally-functional and prepared by inter-polymerising suitable functional and non-functional monomers. Some examples of the former are presented in Table 7.6. It is difficult to obtain low functionalities since the random distribution of functional monomer would lead to a proportion of non-functional molecules or those furnished with only one group.

Amongst functionality which is difficult to introduce directly may be mentioned that which reacts with acryloyl unsaturation (such as primary amine), and that which interferes with the free radical reaction (such as phenolic hydroxyl).

Other chain growth polymers yield terminal functionality, for example the ionic polymerisation of ethylene oxide, propylene oxide, or tetrahydrofuran can lead to hydroxyl-telechelic polyethers whose end groups can be transformed as described above, or converted to primary amines, when they constitute an important class of epoxy crosslinker.

Table 7.6 *Ambifunctional acrylic monomers (R = H or methyl)*

Structure	*Name*	*Reactive group*
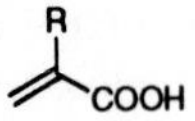 R-substituted acrylic COOH	(Meth)acrylic acid	—COOH
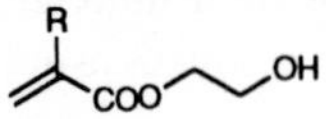 COO–CH₂CH₂–OH	Hydroxyethyl (meth)acrylate	—OH
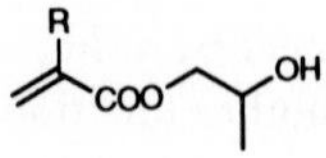 COO–CH₂CH(CH₃)–OH	Hydroxypropyl (meth)acrylate	—OH
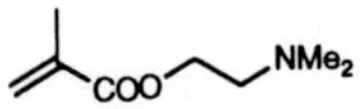 COO–CH₂CH₂–NMe₂	Dimethylaminoethyl methacrylate	—NR_2
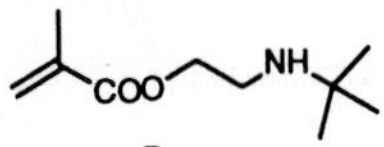 COO–CH₂CH₂–NH–tBu	Tertbutylaminoethyl methacrylate	>NH
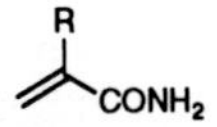 CONH₂	(Meth)acrylamide	—$CONH_2$
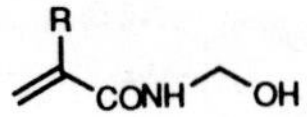 CONH–CH₂–OH	Methylol(meth)acrylamide	activated —OH
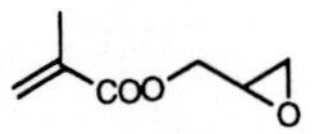 COO–CH₂–epoxide	Glycidyl(meth)acrylate	Epoxy
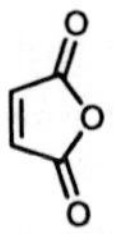 maleic anhydride ring	Maleic anhydride	Cyclic anhydride
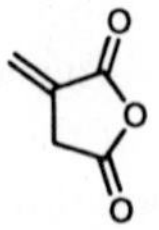 itaconic anhydride ring	Itaconic anhydride	Cyclic anhydride
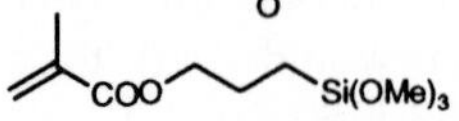 COO–(CH₂)₃–Si(OMe)₃	Trimethoxysilylpropyl methacrylate	Hydrolysable silyl ether
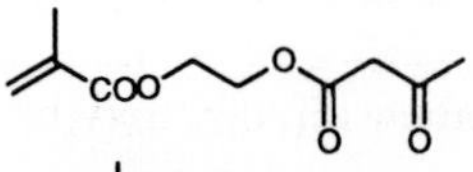 COO–CH₂CH₂–O–CO–CH₂–CO–CH₃	Acetoacetoxyethyl methacrylate	Acidic carbon
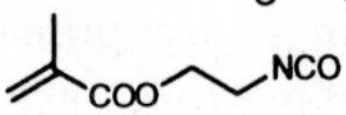 COO–CH₂CH₂–NCO	Isocyanatoethyl methacrylate	Isocyanate
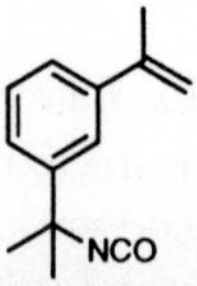 m-isopropenyl-C₆H₄–C(CH₃)₂–NCO	Dimethylisocyanatomethyl-3-isopropenylbenzene	Isocyanate

Chain growth of polycaprolactone from an initiator site (hydroxyl or carboxyl) yields polyesters with terminal groups reflecting the chemistry of the initiator (Scheme 9). It provides a convenient route to telechelic or radiate polymers, blocks, or grafts when diols or polyols or terminally or laterally functional precursors are used.

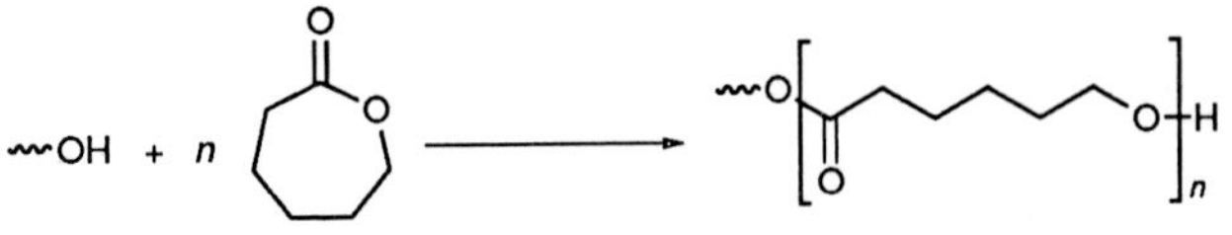

Scheme 9

CROSSLINKING CHEMISTRIES

It is helpful to classify the various crosslinking chemistries by reaction type and conditions of use.

The main distinction is between 'ambient cured' and 'stoved' systems which differ markedly in their chemistry and physics. Condensation chemistries are favoured for stoving systems (with the notable exception of powder coatings). They are less commonly used for ambient cure systems where removal of volatiles from the film may present difficulties. More reactive functional groups are used and a two pack approach is usually adopted. The two reactive components are mixed immediately before use. The high reactivity of such systems is often accompanied by high toxicity and appropriate precautions must be observed.

A convenient classification of the intermolecular chemistries involved is given in Table 7.7. The ring opening processes can be sub-divided according to ring size.

Heterolytic Step Growth Ring Opening

Heterolytic step growth ring opening processes involve formation of a single linkage between two units when an activated ring on one is subject to nucleophilic attack by an active hydrogen containing group on the other. Ring strain is exploited to contribute to the reactivity of the system.

Isocyanate Chemistry (2-membered rings). Isocyanate chemistry was investigated by Bayer in the 1930s with a view to producing polymers with the attractive hydrogen-bonding characteristics of polyamides. It proved that incorporation of urethane groups into polymeric systems produced marked enhancements of their properties, attributed to secondary

Table 7.7 *Classification of crosslinking chemistries*

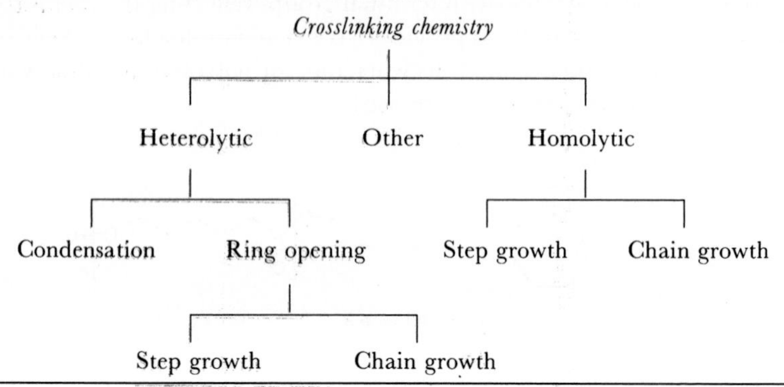

crosslinking through their hydrogen bonds (15) to the extent that a pure polyurethane polymer would be quite intractable, and it was sufficient to incorporate a few urethane linkages per molecule, as when crosslinking.

```
~~N—C—O~~
  |  ||
  H  O
  :  :
  O  H
  || |
~~O—C—N~~
```

(15)

Some of the reactions of isocyanates with active hydrogen compounds are given in Schemes 10–13. At elevated temperatures, the products of the reactions can react with further quantities of isocyanate (Schemes 14–16).

$$\sim\sim N{=}C{=}O + HO\sim\sim \longrightarrow \sim\sim NH{-}\overset{\overset{\displaystyle O}{\|}}{C}{-}O\sim\sim$$

URETHANE

Scheme 10

$$\sim\sim N{=}C{=}O + H_2N\sim\sim \longrightarrow \sim\sim NH{-}\overset{\overset{\displaystyle O}{\|}}{C}{-}NH\sim\sim$$

UREA

Scheme 11

$$\sim\sim N{=}C{=}O + HOOC\sim\sim \longrightarrow \sim\sim\left[NH{-}\overset{\overset{\displaystyle O}{\|}}{C}{-}O{-}\overset{\overset{\displaystyle O}{\|}}{C}\right] \xrightarrow{-CO_2} \sim\sim NH{-}\overset{\overset{\displaystyle O}{\|}}{C}\sim\sim$$

AMIDE

Scheme 12

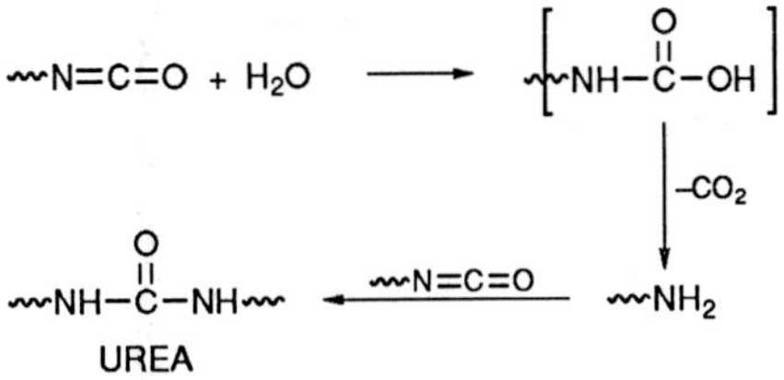

$$\sim N{=}C{=}O + H_2O \longrightarrow [\sim NH{-}\overset{O}{\overset{\|}{C}}{-}OH] \xrightarrow{-CO_2} \sim NH_2$$

$$\sim NH_2 \xrightarrow{\sim N=C=O} \sim NH{-}\overset{O}{\overset{\|}{C}}{-}NH\sim$$

UREA

Scheme 13

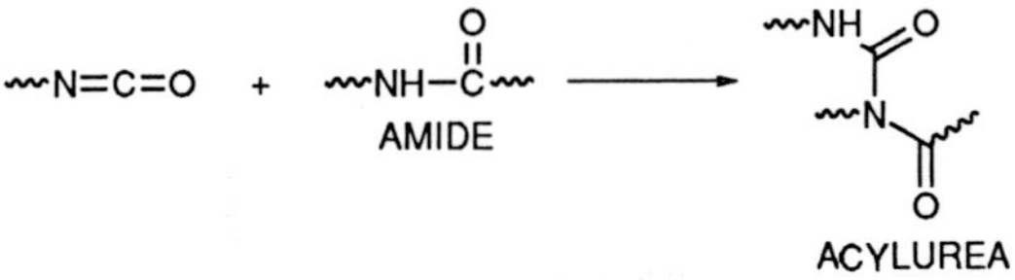

$$\sim N{=}C{=}O + \sim NH{-}\overset{O}{\overset{\|}{C}}\sim \longrightarrow \sim N(\overset{O}{\overset{\|}{C}}{-}NH\sim)(\overset{O}{\overset{\|}{C}}\sim)$$

AMIDE → ACYLUREA

Scheme 14

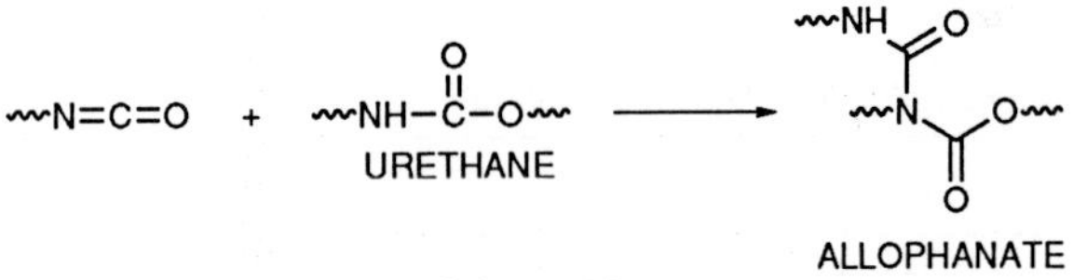

$$\sim N{=}C{=}O + \sim NH{-}\overset{O}{\overset{\|}{C}}{-}O\sim \longrightarrow \sim N(\overset{O}{\overset{\|}{C}}{-}NH\sim)(\overset{O}{\overset{\|}{C}}{-}O\sim)$$

URETHANE → ALLOPHANATE

Scheme 15

$$\sim N{=}C{=}O + \sim NH{-}\overset{O}{\overset{\|}{C}}{-}NH\sim \longrightarrow \sim N(\overset{O}{\overset{\|}{C}}{-}NH\sim)(\overset{O}{\overset{\|}{C}}{-}NH\sim)$$

UREA → BIURET

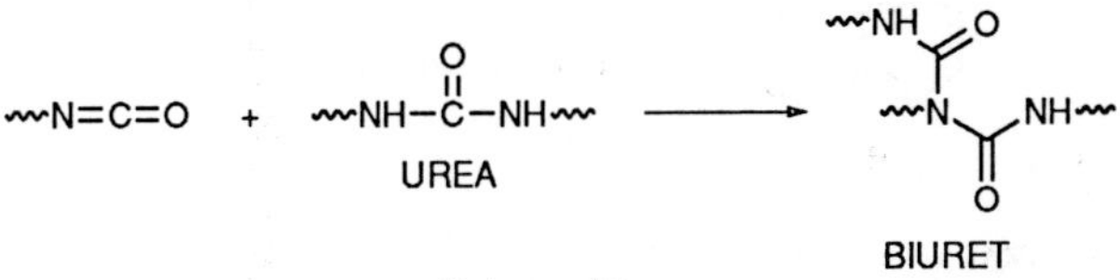

Scheme 16

A further aspect of isocyanate chemistry is the possibility of catalytic dimerisation and trimerisation (Schemes 17 and 18). The dimer exists as reactive four membered rings, used as blocked isocyanates, whereas the six membered trimer is very stable.

A number of difunctional isocyanate monomers are listed in Table 7.8. They are prepared, with the exception of tetramethylxylyenediisocyanate (TMDXDI), by the action of phosgene (carbonyl chloride) on the corresponding diamines. Most isocyanates produce a very marked physiological reaction in certain individuals. It usually manifests itself as a form of respiratory distress known as industrial asthma, whose unfortunate victims become sensitised to very low levels of airborne isocyanate.

Accordingly, great care must be taken when using isocyanate in coatings, where they have the maximum opportunity of entering the atmosphere, either as vapour or aerosol (due to spraying).

$$2 \sim N{=}C{=}O \longrightarrow$$

URETDIONE

Scheme 17

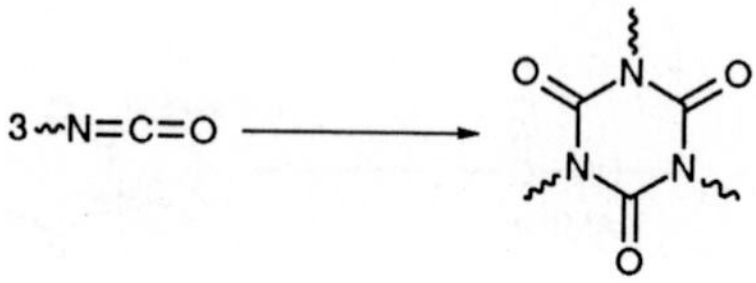

$$3 \sim N{=}C{=}O \longrightarrow$$

ISOCYANURATE

Scheme 18

Table 7.8 *Some common difunctional isocyanates*

Structure	*Name*
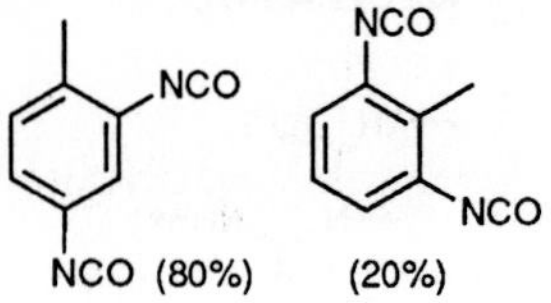 NCO (80%) (20%)	Tolylenediisocyanate (TDI)
OCN—$(CH_2)_6$—NCO	Hexamethylenediisocyanate (HDI)
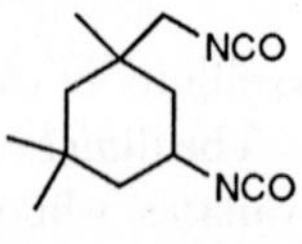 NCO NCO	Isophoronediisocyanate (IPDI)
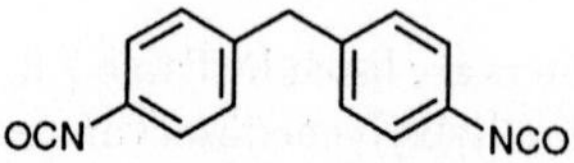 OCN NCO	Methylenediisocyanate (MDI)
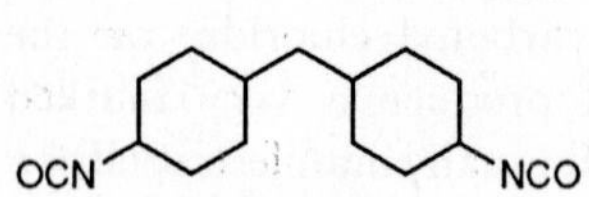 OCN NCO	Hydrogenated MDI (HMDI)
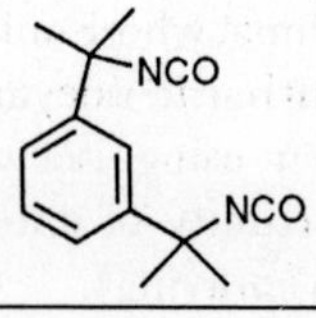 NCO NCO	*m*-Tetramethylxylylenediisocyanate (TMXDI)

Diphenylmethanediisocyanate (MDI) and its hydrogenated equivalent (HMDI) find limited use in coatings compositions being of moderate volatility and toxicity, *m*-TMXDI is a recent addition to the armoury which is claimed to be free of respiratory sensitisation problems. Tolylenediisocyanate (TDI) and hexanediisocyanate (HDI), however, are the most widely used in coatings curing agents, and both are powerful sensitisers and sufficiently volatile to constitute a hazard. It is therefore necessary to use them in 'prepolymer' form which has the advantage of increasing their functionality to three. TDI is usually adducted 3: 1 with a triol such as trimethylolpropane. HDI is either reacted 3: 1 with water to produce a biuret (16) or catalytically trimerised to the isocyanurate (17), which is generally preferred for its hydrolytic stability and additional 'hardness'. Addition polymers from the unsaturated isocyanates noted in Table 7.8 are likely to become increasingly important. HDI prepolymers are used to crosslink hydroxyacrylics or hydroxypolyesters at ambient temperature to films of excellent weathering resistance, and application properties. They are used as top coats, for example in car refinish and aircraft coating.

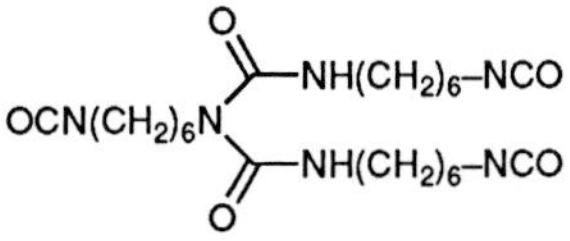

$OCN(CH_2)_6N[C(=O)NH(CH_2)_6NCO]_2$

(16)

Isocyanurate ring with three $(CH_2)_6NCO$ substituents on N

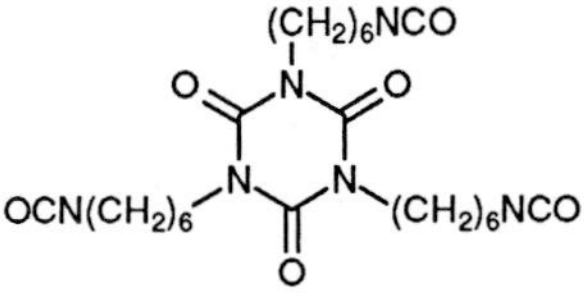

(17)

The reaction of isocyanates with hydroxyl compounds is sluggish in the absence of catalyst and various organotin compounds and tertiary amines are used. The addition of acetylacetone appears to moderate reactivity and can be used to extend pot life without impairing the properties of the final film.

Another interesting approach to 'cure latency' involves the use of tertiary amine vapour as catalyst. It is introduced with the spray propellant, or the newly coated article is passed through a vapour-filled chamber and very rapid curing takes place.

Moisture curing of isocyanates is somewhat limited in its application since the evolution of carbon dioxide is likely to disrupt a thick film. However, it has the advantage of being 'latent'—the liquid paint is stable indefinitely but cures rapidly as a film on contact with atmospheric moisture.

Curing with primary and secondary amines is finding increasing favour because the urea linkage generated is even more hydrolytically stable than urethane, and because the reaction is extremely rapid. No catalysis is required and it is usually necessary to restrain the process by derivatising the amine as ketimine, enamine, or oxazolidine which release amine only after hydrolysis by atmospheric moisture (Schemes 19–21).

Aromatic amines are less reactive and have been used without protection in elastomeric compositions.

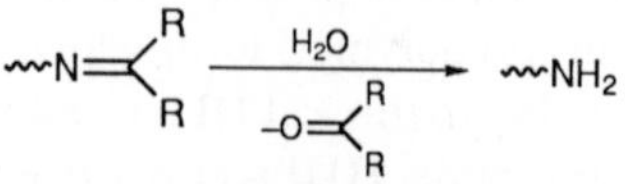

Scheme 19

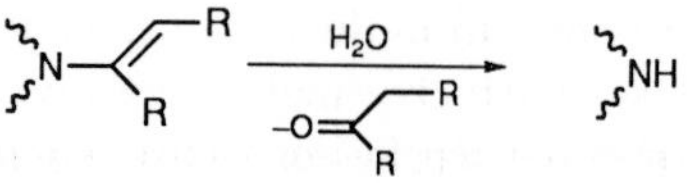

Scheme 20

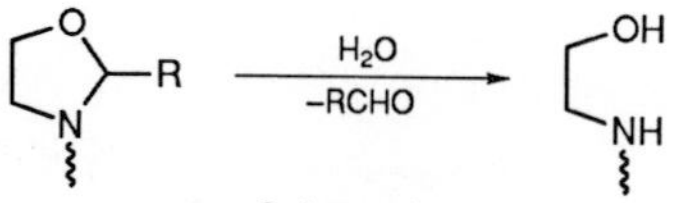

Scheme 21

The extensive chemistry of 'blocked' isocyanates is considered under condensation processes.

Activated Olefin Chemistry (2-membered ring). Curing chemistry based on Michael and Michael type reactions has increasingly appeared in the patent literature in recent years and at least one technology is now commercial.

It involves addition of an activated hydrogen species to an activated olefin such as an acrylate (Scheme 22).

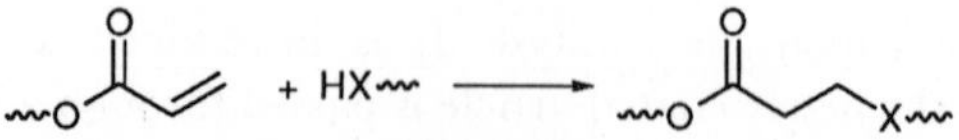

Scheme 22

In the true Michael reaction, X is an acidic carbon, and an alkali metal alkoxide or a strong organic base such as an amidine is required as catalyst. The related reactions where X is the residue of primary or secondary amines are self catalysing; thiols require a tertiary amine catalyst. Proposed technologies use high and low molecular weight acrylate esters, maleate or itaconate polyesters, or bismaleimides as acceptors, and acetoacetate functional acrylics, malonate polyesters, polythiols, polyamines (optionally protected as ketimines), and even polyols as donors.

Such systems react readily at ambient temperature and have been proposed for car refinishing and similar coating operations, though an acidic substrate can inhibit cure.

It is also possible to protect the acrylate moiety as an adduct of monomeric amine. On stoving with a polymeric amine, the volatile material is lost and curing ensues.

Carbodiimide Chemistry (2-membered ring). Another proprietary innovation suggested for use with water-based systems is the ambient temperature interaction of carbodiimides with carboxylic acids to yield acylureas (Scheme 23). Polyfunctional carbodiimides are prepared by the condensation of poly- and monofunctional isocyanates, or dehydration of suitable ureas.

~N=C=N~ + HOOC~ ⟶ ~NH–C(=O)–N(~)–C(=O)~

Scheme 23

Epoxide Chemistry (3-membered ring). The epoxide (oxirane) ring is highly reactive towards a number of nucleophilic species (Scheme 24). Any molecule containing two or more oxirane groups can be described as an 'epoxy', but the term epoxy resin is usually understood to denote the oligomers obtained by condensation of 'bisphenol A' or similar polyphenol with epichlorohydrin (Scheme 25). Bisphenol A itself is prepared by condensing phenol with acetone.

The $n = 0$ species (18) is liquid and is used in its own right for high and 100% non-volatile coatings. It is also used as a feedstock to prepare solid grades ($n = 1$–4) by 'advancement' with additional bisphenol A. One manufacturer (Shell) however produces the solid grades directly from bisphenol A and epichlorohydrin in appropriate proportions. The latter procedure produces twice as many oligomers in a similar molecular weight envelope.

Despite the problem common to step growth polymers that their functionality will always fall somewhat short of the ideal level of two, epichlorohydrin–bisphenol A polymers yield coatings characterised by hardness, chemical resistance, and excellent adhesion. Unfortunately their weathering resistance is very poor and exterior exposure leads rapidly to a condition known as 'chalking'—the polymer is eroded and powdery pigment is left on the surface.

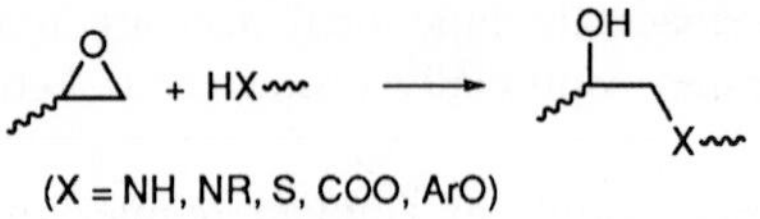

(X = NH, NR, S, COO, ArO)

Scheme 24

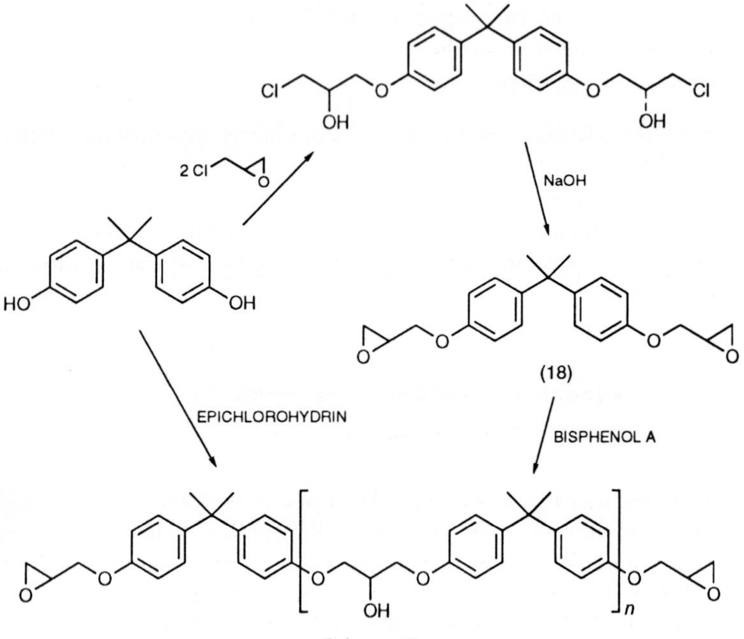

(18)

EPICHLOROHYDRIN

NaOH

BISPHENOL A

Scheme 25

Amongst other available aromatic epoxies are epoxidised 'novolacs' (19) derived from phenol formaldehyde products, and epoxidised amino phenols (20). They offer increased functionality and, consequently, hardness in the cured film.

Improved weathering performance can be obtained by using aliphatic epoxies, though usually at the expense of hardness. Amongst them are hydrogenated versions of liquid bisphenol A epoxide, copolymers of glycidyl methacrylate and related monomers (*c.f.* Table 7.6), epoxidised oils or polybutadienes, and materials such as (21), (22), and (23). However, 'internal' epoxy groups reduced reactivity towards nucleophiles and are more susceptible to self condensation induced by Lewis or Brönsted acids. An epoxy functional melamine–formaldehyde condensate containing amide linkages has recently been commercialised, and claims highly reactive epoxide groups.

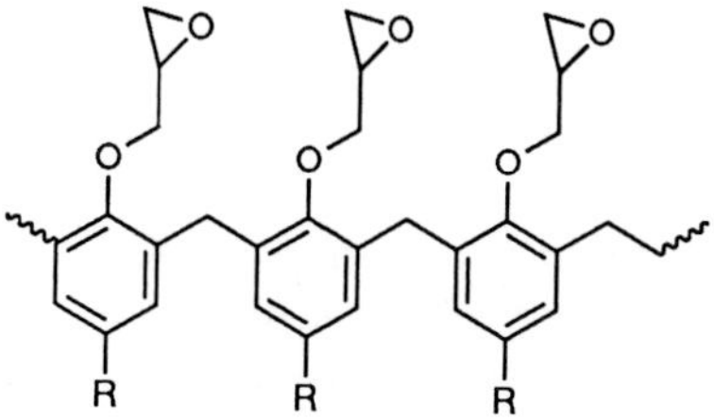

(19) 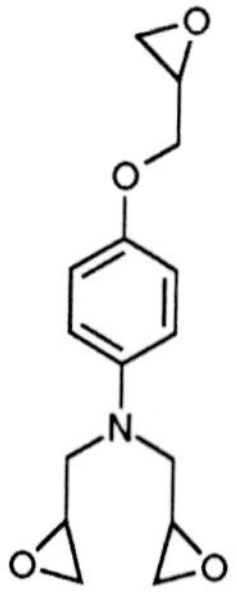 (20)

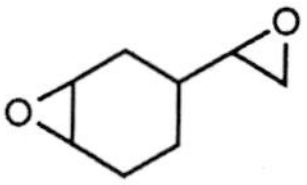

(21) 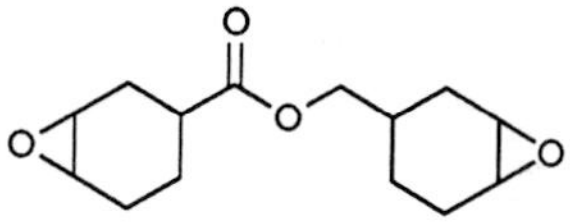 (22)

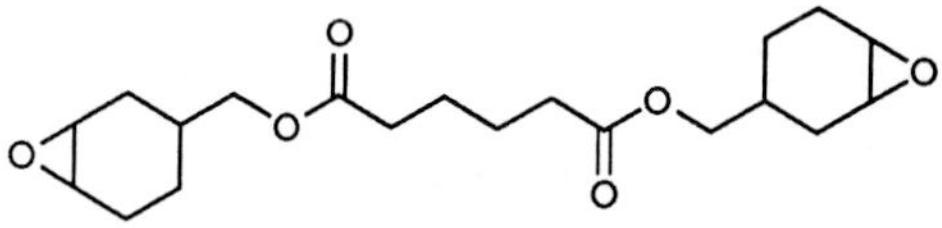

(23)

For ambient temperatures, amines are the curing agents of choice. Simple aliphatic polyamines such as ethylenediamine, diethylenetriamine, and triethylenetetramine are used. All active hydrogens participate so that diethylenetriamine (24), for example, is pentafunctional, but primary amines are more reactive than secondary. The generated tertiary amines catalyse the reaction of epoxies with adventitious hydroxyl groups.

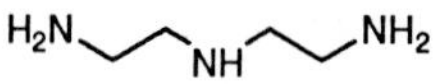

H_2N NH NH_2

(24)

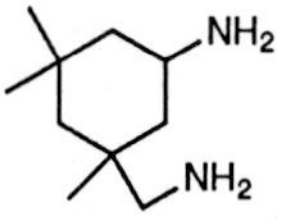

NH_2 NH_2

(25)

Simple amines are rather volatile and toxic and, being strongly basic, tend to react with carbon dioxide at the coating/air interface with consequent cure inhibition and surface stickiness. Cycloaliphatic amines such as isophoronediamine (25), amine tipped polyethers, and prepolymers prepared from simple polyamines and epoxide resins are all used to address the above problems.

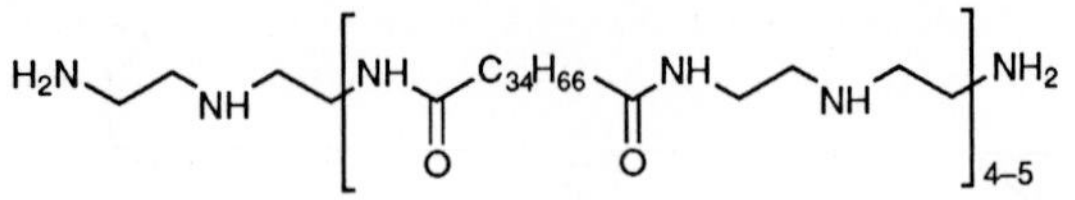

(26)

Amino functional polyamides formed from simple amines and long chain diacids including dimerised fatty acids (*c.f.* Table 7.1) (*e.g.* 26) are effective epoxy curing agents. They can be made to contain various levels of imidazoline by dehydration of chain segments to varying extents (Scheme 26). High levels of imidazoline give lower reactivity since some amine groups are latent. Polyamides provide highly flexible and corrosion resistant thermosets. Mannich bases such as those derived from amine, phenol, and formaldehyde (Scheme 27) show greatly enhanced reactivity, apparently owing to the catalytic effect of the phenolic hydroxyl.

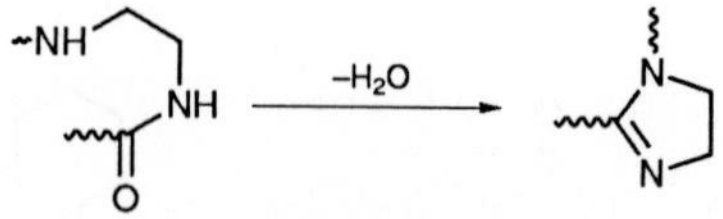

Scheme 26

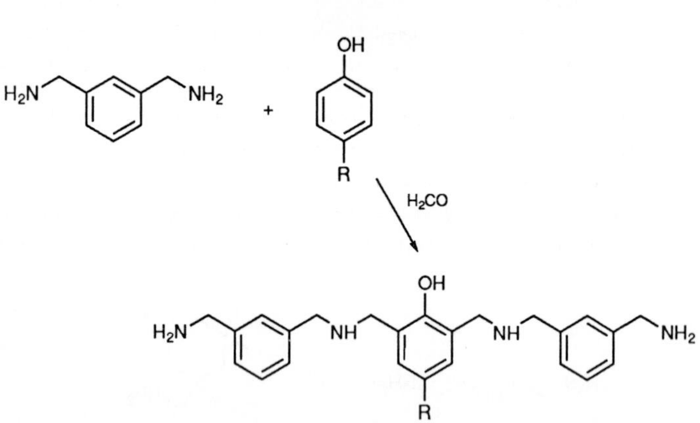

Scheme 27

It is also, of course, possible to present amines in protected form, such as ketimine, so as to provide a latent cure dependent on atmospheric moisture.

Various aromatic polyamines are effective curing agents at stoving temperatures but they are generally rather more toxic than aliphatic amines so find little application in coatings.

Amine cured, aromatic epoxy coatings are used in large amounts as primers for all sorts of metal structures from bridges to aircraft, where their chemical resistance and adhesive qualities are of value, and their poor weatherability does not present a problem.

A number of commercial technologies use aliphatic epoxy amine systems as top coats. In one example a glycidyl methacrylate copolymer is crosslinked with a primary amine functional acrylic, presumably made by reacting an acid functional polymer with ethyleneimine (Scheme 28).

COOH + H–N (aziridine) ⟶ COO–CH₂CH₂–NH₂

Scheme 28

Epoxythiol chemistry has not been much exploited in coatings, perhaps because of the malodorous character of most thiols. However, technically excellent thermosets free of carbonation problems can be prepared using suitable polythiols and tertiary amine catalysts.

Reactions between epoxies and carboxylic acids are relatively sluggish at ambient temperature, and polyacid curing agents find their most important application at elevated temperature in powder coatings (*q.v.*). That said, a number of successful proprietary chemistries achieve ambient curing of aliphatic epoxies with suitable acid functional systems.

Aziridine Chemistry (3-membered ring). The reaction of polycarboxylic acids with polyfunctional aziridines (ethyleneimines) (Scheme 29) proceeds readily at ambient temperature and has been used widely in water-based systems. It is however declining in popularity because of the mutagenic activity of aziridines.

~~N◁ + HOOC~~ ⟶ ~~NH⌒⌒OOC~~

Scheme 29

Anhydride Chemistry (5-membered ring). The two trigonal carbons in a 5-membered cyclic anhydride provide sufficient ring strain for high reactivity (Scheme 30). As expected, amines are very much more reactive than aliphatic alcohols, which require a tertiary amine catalyst for reaction at ambient temperature; phenols and thiols hardly react at all. β-Hydroxyamides are used in certain stoving systems being more reactive with anhydride polymers than straightforward polyols, and their reactivity has been attributed to thermal rearrangement to aminoester (Scheme 31).

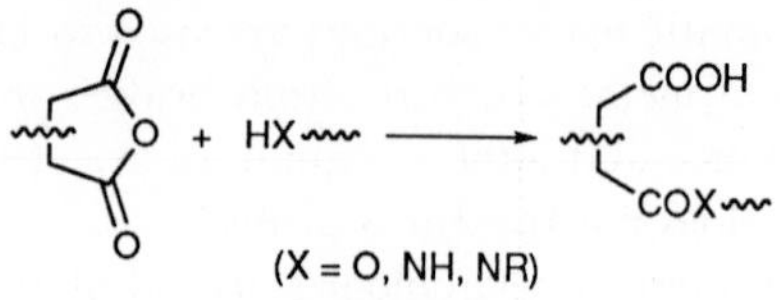

(X = O, NH, NR)

Scheme 30

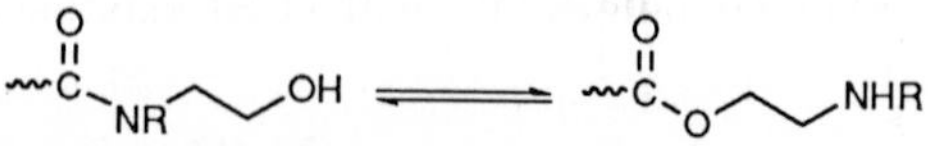

Scheme 31

Suitable anhydride functional polymers can be obtained by copolymerising maleic or itaconic anhydride (see Table 7.6) with hydrocarbon or acrylic monomers or, with rather more difficulty, by incorporating trimellitic anhydride (see Table 7.1) into a polyester.

Other Ring Opening Chemistry. Other ring opening processes which have received attention in recent years are as Schemes 32–34, involving cyclic carbonates with amines, azlactones with amines or alcohols, and oxazolines with acids. A six membered ring, oxadiazahexatrione derived from diols, isocyanates, and carbon dioxide is said to crosslink with hydroxyl groups in the same molecule on exposure to amine vapour. It possibly behaves as an anhydride (Scheme 35).

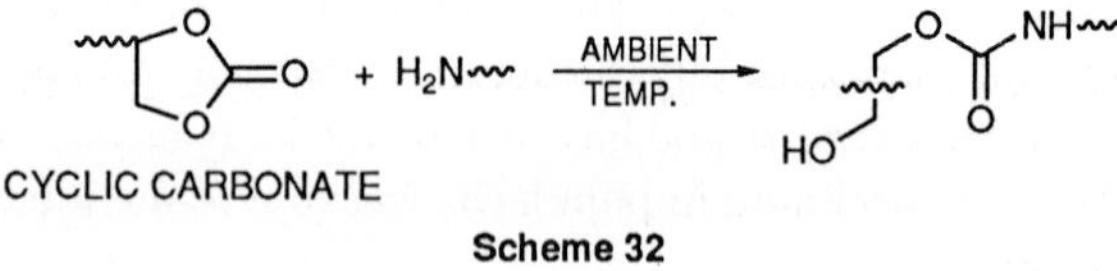

CYCLIC CARBONATE $+ H_2N$~ $\xrightarrow[\text{TEMP.}]{\text{AMBIENT}}$

Scheme 32

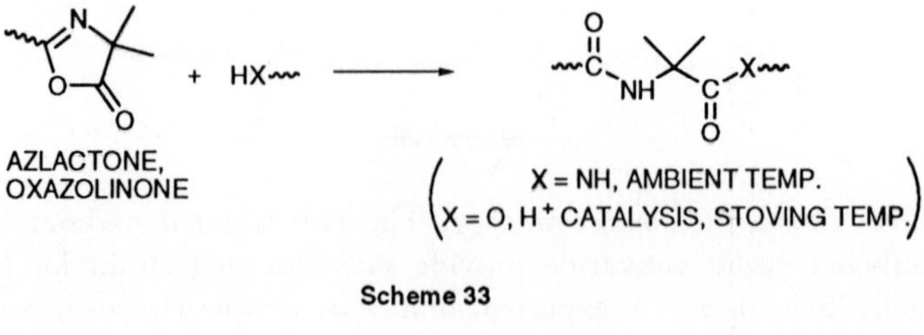

AZLACTONE, OXAZOLINONE + HX~ ⟶

(X = NH, AMBIENT TEMP.
X = O, H^+ CATALYSIS, STOVING TEMP.)

Scheme 33

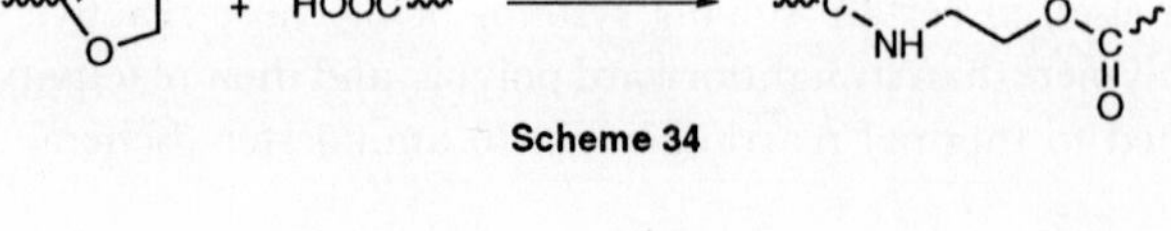

+ HOOC~ $\xrightarrow{130\,^{\circ}C}$

Scheme 34

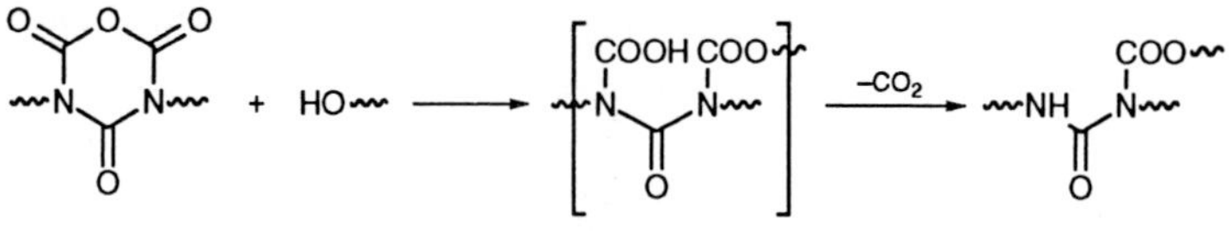
COOH COO
COO
HO
–CO2
NH

Scheme 35

Heterolytic Chain Growth Ring Opening

As with chain growth polymerisation, chain growth crosslinking does not involve discrete steps but a process which will continue to consume reactive groups until it is stopped. Each functional group provides two linkages to another polymer molecule hence relatively dense crosslinking is obtained (Scheme 36). However, physical constraints, such as the increasing immobility of available functional groups, mean that chains are likely to be very short. The development of ionic features or unsaturation at the chain ends should also be taken into account when assessing the performance characteristics of such materials.

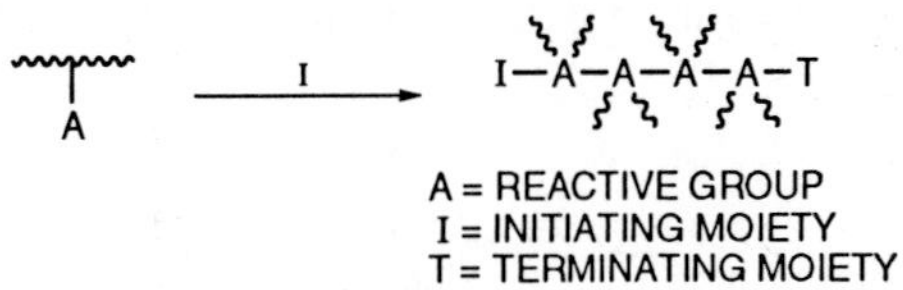
A
I
I—A—A—A—A—T
A = REACTIVE GROUP
I = INITIATING MOIETY
T = TERMINATING MOIETY

Scheme 36

Epoxy Self Addition (3-membered rings). Epoxy polymers, especially those containing internal epoxide groups can be induced to undergo a cationic polymerisation by treatment with strong acids such as BF_3, $SnCl_6$, H_2PF_6, or CF_3SO_3H (Scheme 37).

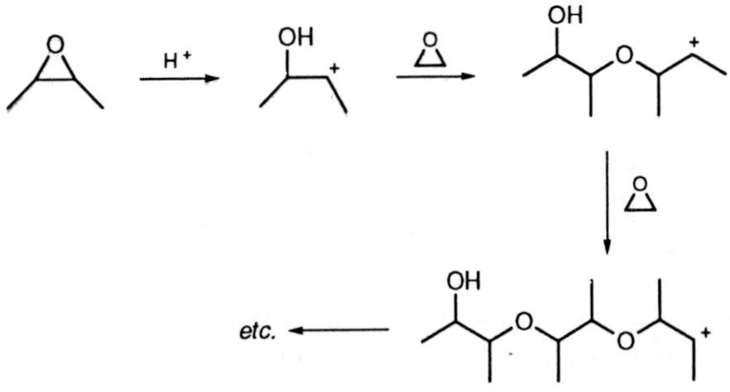
H+
OH
etc.

Scheme 37

The acid can be generated photochemically, providing efficient photo-curing systems. Early technologies used diazonium salts of strong acids which became effective when the diazonium cation decomposed under ultraviolet irradiation. More recently, sulfonium salts have been preferred (Scheme 38).

$$Ar_3S^+ \ PF_6^- \xrightarrow{h\nu} H^+ \ PF_6^-$$

Scheme 38

Allyl ether functional systems can be cationically cured in an analogous way.

Epoxy/Anhydride Alternating Addition (3-membered rings). Anhydride and epoxy residues have a strong tendency to undergo alternating chain growth which greatly exceeds the affinity of carboxylic acids for epoxides or hydroxyl groups for anhydrides, though all four functional groups are frequently formulated together. The process can be initiated by a tertiary amine and the mechanism of Scheme 39 has been proposed.

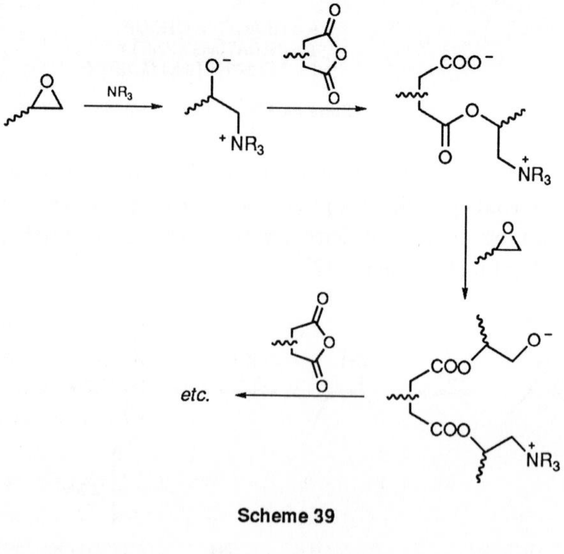

NR_3 O^- $^+NR_3$ COO^- $\overset{+}{N}R_3$ etc. COO COO O^- $\overset{+}{N}R_3$

Scheme 39

The technology has been most frequently used in powder coatings where solid bisphenol A epoxy or a glycidyl methacrylate copolymer is cured with monomeric anhydride such as phthalic or trimellitic, or a maleic anhydride polymer.

Condensation Processes

Condensation crosslinking involves formation of a crosslink between two functional groups with the elimination of a small molecule such as water or alcohol. It is frequently a satisfactory mechanism for stoved coatings when the volatile entity can be removed in a controlled way, but works less well in ambient cured systems.

Phenol Formaldehyde Polymers and Crosslinkers. Amongst the oldest synthetic polymer materials, 'phenolics' are prepared by condensing phenol and formaldehyde. Phenol is reactive at the 2, 4, and 6 positions and will condense with formaldehyde ultimately forming a densely crosslinked network known as 'Bakelite'. In the coatings industry it is more usual to employ 4-substituted phenols so that linear materials are produced (Scheme 40).

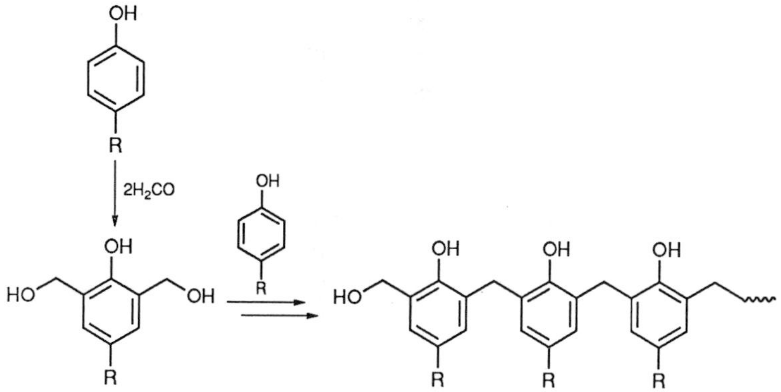

OH R $2H_2CO$ OH HO OH R OH R HO OH OH OH R R R

Scheme 40

Use of excess formaldehyde leads to short chains terminated with methylol groups (as shown) and known as 'resoles'. On the other hand, an excess of phenol provides 'novolacs' with reactive ring hydrogens at the chain ends. In-chain reactivity can be introduced by judicious incorporation of unsubstituted phenol.

Novolacs will condense with further formaldehyde or resoles, but otherwise are inert except with respect to their phenolic groups which can be used to cure epoxy polymers or converted to epoxy groups. The reactive groups on resoles are reactive with active hydrogen species on stoving under basic, acidic, or neutral conditions, though acidic catalysis is most effective. The reaction proceeds via a quinomethide intermediate (Scheme 41).

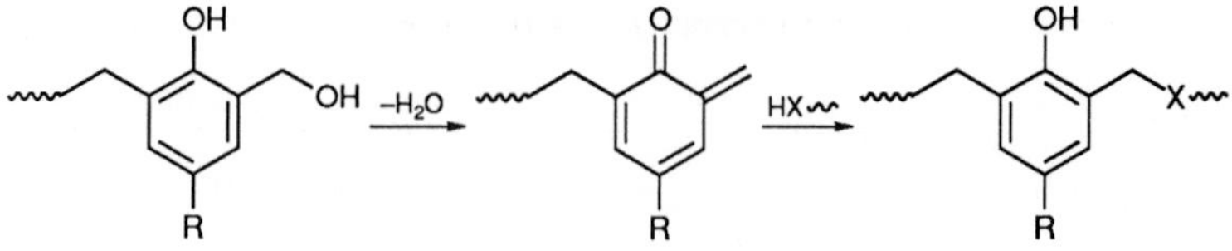

Scheme 41

The dark colour associated with phenolic polymers prevents their widespread use in coatings but they are valued, for example, in can linings for their great chemical inertness, and are often used in conjunction with epoxies.

Amino Formaldehyde Crosslinkers and Polymers. The most important chemistry used in stoved coatings is based on optionally alkylated, amino-formaldehyde condensates. The principal 'amino' precursors are urea (27), thiourea (28), melamine (29), benzoguanamine (30), glycoluril (31), and copolymers of acrylamide (32).

(27) (28) (29) 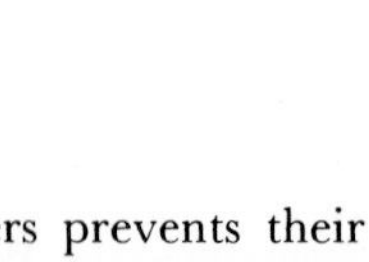 (30)

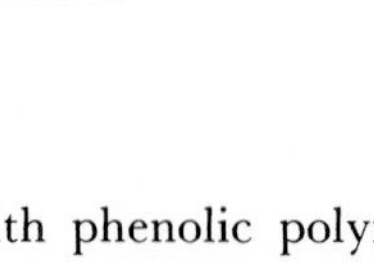

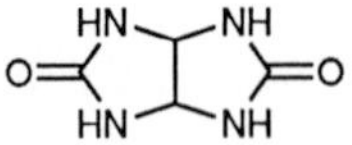

(31) 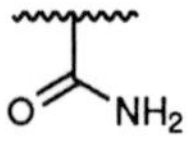 (32)

Melamine–formaldehyde products are the most widely used. By comparison with melamine, urea gives highly reactive polymers with rather poor water resistance. They can be cured at room temperature using a strong acid catalyst and are used in furniture finishes combined with hydroxyl functional polyesters. Benzoguanamine gives increased flexibility but reduced solvent and weathering resistance. Glycoluril resins emit relatively little formaldehyde and are said to be useful in high solids coatings. Acrylamide polymers naturally reflect the properties of the co-monomers, but can be cured at low temperatures and are used, for example, in domestic appliance finishes where good colour and alkali resistance are important.

The crosslinking chemistry can be illustrated by reference to melamine derivatives.

Fully methylolated melamine–formaldehyde (33) is an essentially discrete substance and is commercially available. It and similar materials require strong acid catalysis to undergo a relatively straightforward condensation with hydroxyl (or carboxyl) functional film formers (Scheme 42). Incompletely alkylated versions react similarly, but with loss of water. However, when significant amounts of –NH remain in the molecule, organic or other weak acids provide sufficient catalysis.

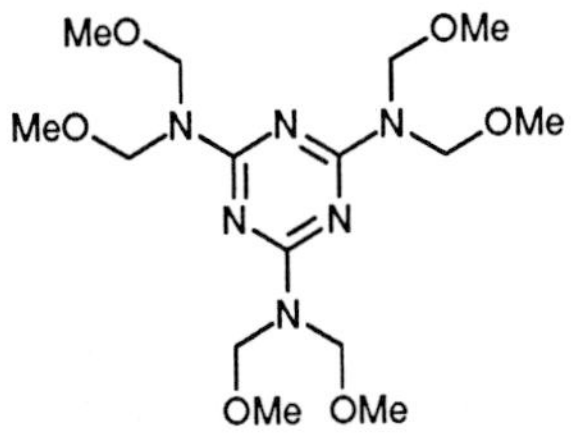

(33)

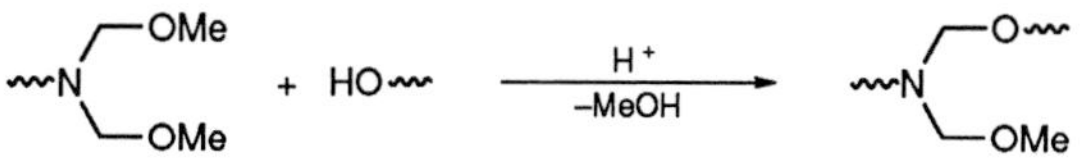

Scheme 42

A much wider range of reactions, including self condensation, becomes possible and it follows that the species, as prepared, will tend to be oligomeric. Continued condensation in the absence of other materials leads to densely crosslinked thermosets which are used as surface coatings for laminates. Reaction with active hydrogen compounds is believed to involve Schiff's base type intermediates (Scheme 43).

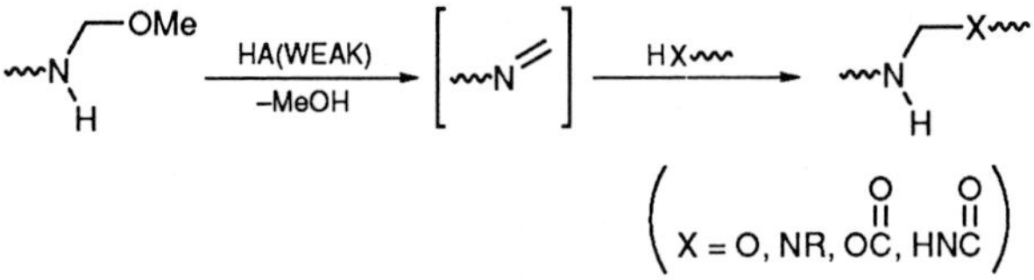

(X = O, NR, OC(=O), HNC(=O))

Scheme 43

Both types of system yield some self-condensed material containing (34) and (35) type bridges and it is usual to formulate coatings to contain considerably more melamine–formaldehyde (MF) than suggested by

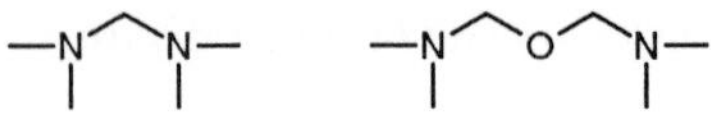

(34) (35)

stoicheiometry. MF crosslinkers are used with polyester and acrylic polyols as high quality stoving top-coats for motor cars.

Blocked Isocyanates. The bond formed from certain mono-active hydrogen compounds and isocyanates is labile, and the so-called 'blocking agent' can be displaced by a polyol or polyamine forming a crosslinked network. The process frequently appears to be a mixture of deblocking followed by addition, and S_N2 displacement, as shown in Scheme 44.

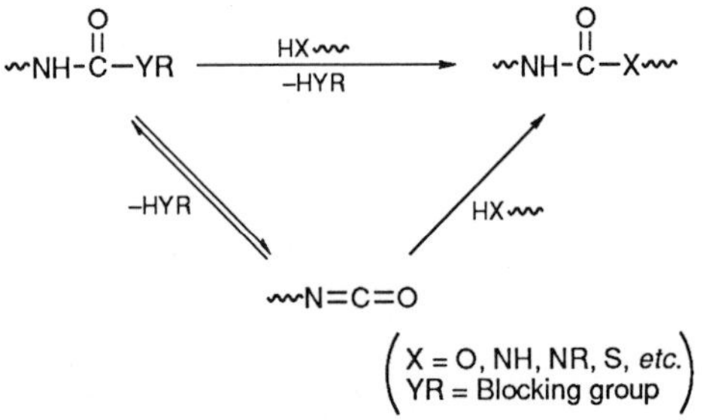

(X = O, NH, NR, S, *etc.*
YR = Blocking group)

Scheme 44

Its rate at a given temperature is influenced by the nature of the attacking species and the blocking group. At one extreme, a phenol-blocked isocyanate can be cured at ambient temperature with a polyamine in the presence of amidine catalyst. However, blocked isocyanates are usually formulated as more or less stable single package compositions which cure on stoving. Alcohol-blocked isocyanate curing agents, for example can be used in cathodic electrodeposition formulations.

Some blocking agents and temperatures at which their aliphatic isocyanate derivatives react rapidly with polyols are listed in Table 7.9. A number of proprietary blocking agents are said to provide crosslinkers effective at lower temperatures, dibutylglycolamide, for example, has been reported to crosslink polyols at 120 °C, but the stability of such a composition at ambient temperature must be in some doubt. Malonic ester has been regarded as a useful isocyanate blocking agent, but recent investigations established that the acylated malonate was undergoing transesterification with polyol by an E_1CB mechanism involving a ketene intermediate (Scheme 45).

Table 7.9 *Effective curing temperatures of polyols with blocked isocyanates*

Blocking agent	*Effective cure temperature*
Aliphatic alcohol	>180 °C
Phenol	175–180 °C
Glycol ether	160 °C
Caprolactam	150–160 °C
Methylethyl ketoxime	150 °C

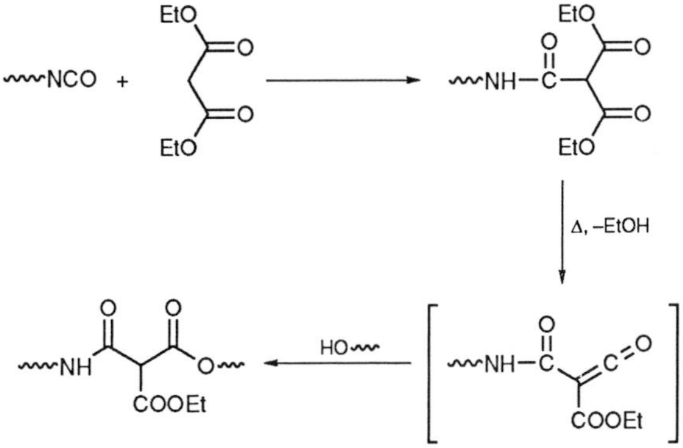

Scheme 45

The possibility of obtaining 'blocked isocyanates' without ever handling free isocyanates has undoubted appeal, and a number of technologies have been proposed.

The reaction of cyclic carbonates with primary amines to produce hydroxyethylurethanes was mentioned under heterolytic step growth ring opening. The products derived from ethylene or propylene carbonate prove to have reactivity comparable with that of phenol blocked isocyanates (Scheme 46).

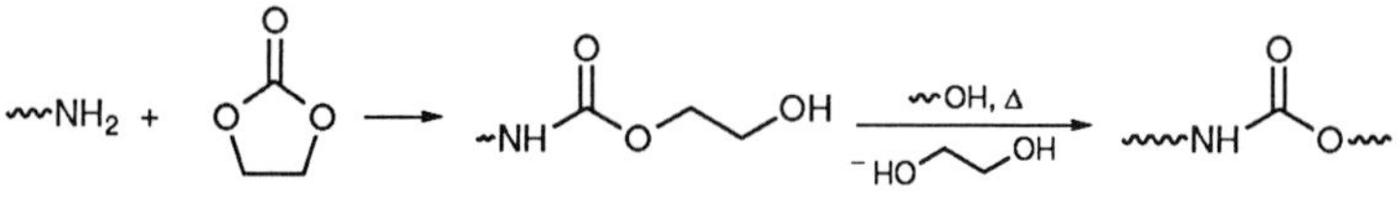

Scheme 46

A hydroxamic ester can undergo a Lossen rearrangement to isocyanate at the required rate at temperatures as low as 100 °C (Scheme 47). They are normally used in two pack compositions.

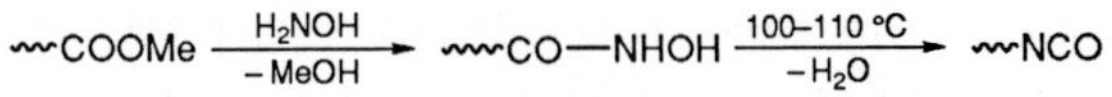

$$\sim\!\!\sim COOMe \xrightarrow[-MeOH]{H_2NOH} \sim\!\!\sim CO{-}NHOH \xrightarrow[-H_2O]{100\text{–}110\,^{\circ}C} \sim\!\!\sim NCO$$

Scheme 47

Aminimides prepared from esters and *N*,*N*-disubstituted hydrazines eliminate tertiary amine with formation of isocyanates (Scheme 48).

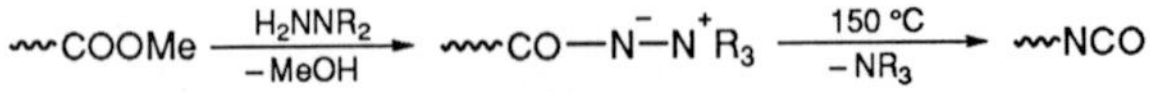

$$\sim\!\!\sim COOMe \xrightarrow[-MeOH]{H_2NNR_2} \sim\!\!\sim CO{-}\bar{N}{-}\overset{+}{N}R_3 \xrightarrow[-NR_3]{150\,^{\circ}C} \sim\!\!\sim NCO$$

Scheme 48

Though not strictly condensation processes, it is convenient to mention the cleavage of bicyclic furoxan species (Scheme 49), the ring opening of bis-cyclic ureas (Scheme 50), and the degradation of low molecular weight polymers containing uretidinedione groups (Scheme 51) to yield diisocyanates.

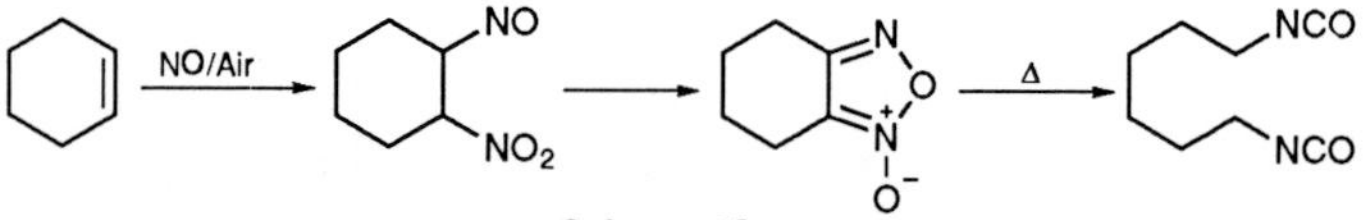

NO/Air; NO; NO_2; Δ; NCO; NCO

Scheme 49

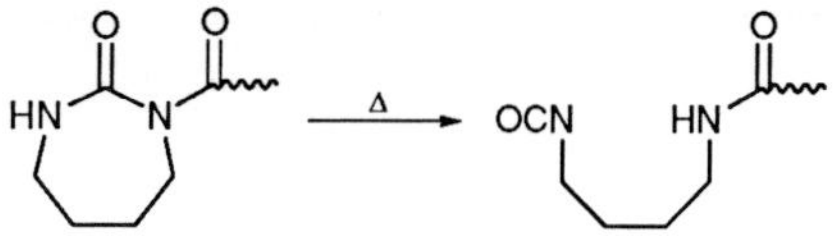

Δ; HN; N; OCN; HN

Scheme 50

$\sim\!\!\sim N$ (uretidinedione) $N \sim\!\!\sim \longrightarrow 2 \sim\!\!\sim NCO$

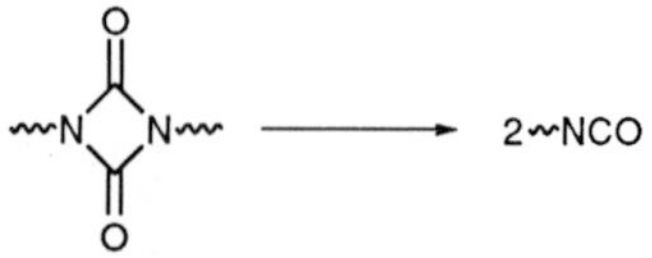

Scheme 51

Silanol Chemistry. The dehydration of silanols to form silicates is the basis of much ceramic sol–gel technology (Scheme 52).

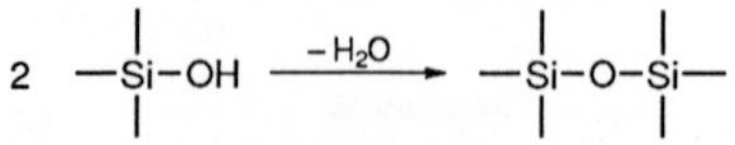

$$2\ {-}\overset{|}{\underset{|}{Si}}{-}OH \xrightarrow{-H_2O} {-}\overset{|}{\underset{|}{Si}}{-}O{-}\overset{|}{\underset{|}{Si}}{-}$$

Scheme 52

It occurs readily on removing water at ambient temperature. Silicate esters provide reasonably stable precursors in the absence of water, but exposure to moist air permits sequential hydrolysis and dehydration until a silica network is obtained. The use of partially condensed tetraethyl silicate as a binder for 'zinc rich' anticorrosive primers is an example (Scheme 53).

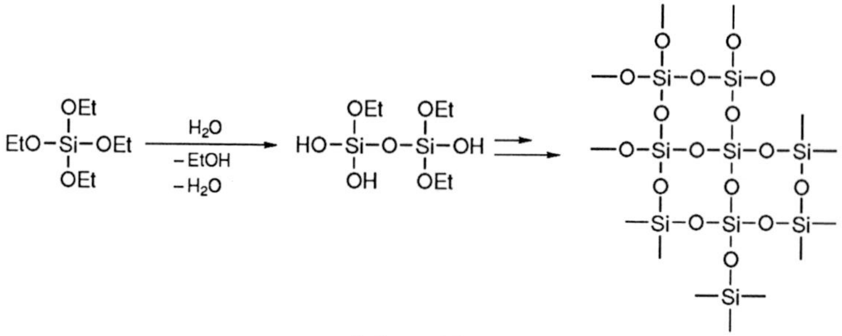

$$\mathrm{Si(OEt)_4} \xrightarrow[-\mathrm{EtOH},\ -\mathrm{H_2O}]{\mathrm{H_2O}} \mathrm{HO{-}Si(OEt)(OH){-}O{-}Si(OEt)_2{-}OH} \rightleftharpoons \longrightarrow \text{silica network}$$

Scheme 53

Such networks are unfortunately extremely brittle, and their formation involves large shrinkage, so films of good integrity are rarely formed, even from very heavily filled compositions. Hybrid organic–inorganic systems have been prepared using such molecules as (36), (37), and (38) to introduce silicate condensation chemistry to an organic backbone.

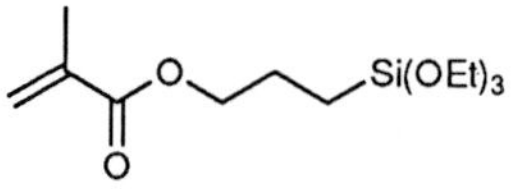

$CH_2{=}C(CH_3)C(O)O(CH_2)_3Si(OEt)_3$

(36)

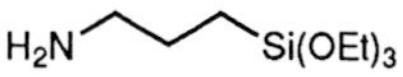

$H_2N(CH_2)_3Si(OEt)_3$

(37)

$OCN(CH_2)_3Si(OEt)_3$

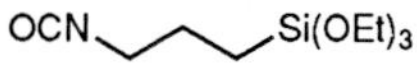

(38)

Silicate condensation is also the basis of 'room temperature vulcanising' silicone rubbers [*e.g.* (39)]. Silicone polymers are highly flexible, soft, weatherable materials with very low surface energy. They can be advantageously combined with silicates or other coatings polymers.

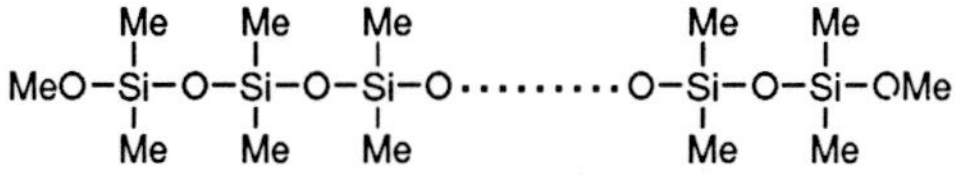

$$\mathrm{MeO{-}SiMe_2{-}O{-}SiMe_2{-}O{-}SiMe_2{-}O}\cdots\cdots\mathrm{O{-}SiMe_2{-}O{-}SiMe_2{-}OMe}$$

(39)

Other Condensation Chemistry. Many crosslinking agents containing other activated ethers or esters, from which a small molecule can be displaced by a nucleophilic group, have been described. An acrylic monomer containing both activated ester and other groups, (methyl methacrylamidoglycolate, methyl ether) is commercially available. Its copolymers are said to react with amines at ambient temperature, or alcohols at elevated temperature (Scheme 54).

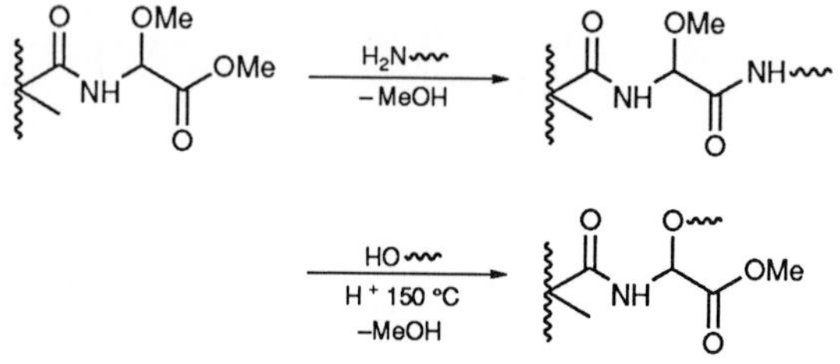

Scheme 54

Step Growth Homolyses

Auto-oxidative Curing. Auto-oxidative crosslinking of allylically unsaturated products is probably the oldest curing chemistry as well as the most widely used. Its mechanism is complex, but Scheme 55 gives an approximate picture.

Adventitious radicals react with singlet oxygen to produce peroxide radicals. The peroxide radical abstracts H from an allylic site generating hydroperoxide and a new carbon radical. The hydroperoxide can be reduced by a transition metal in a low oxidation state to an oxide radical which in turn generates a further alkyl radical. A steady supply of radicals isomerises and couples and with suitable molecular architecture can give rise to a cured product.

Chinese lacquer, the refined sap of a small tree, appears to have been in use in the Orient some 5000 years ago. It is a complex mixture containing urushiol (40) and related materials. Auto-oxidative processes must contribute to the cure, though enzyme promoted oxidative coupling of the phenolic residues also appear to play a part in the formation of coatings of (so far) unsurpassed endurance.

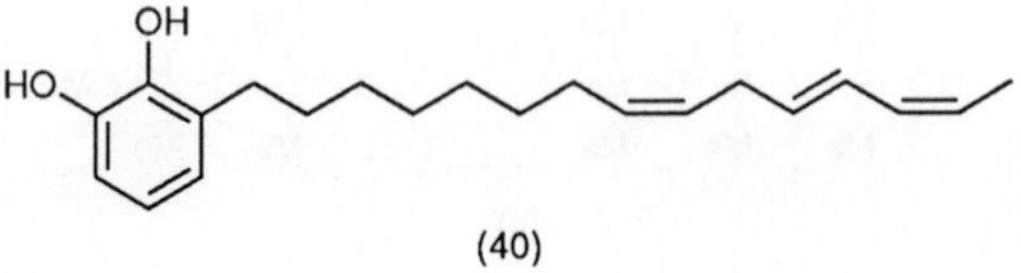

(40)

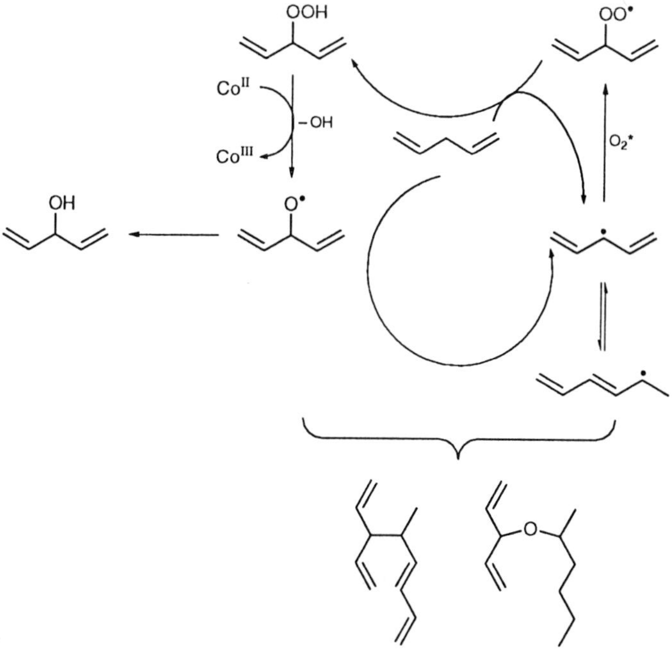

Scheme 55

The exploitation of linseed oil in coatings may have begun as early as Roman times, and since then many other 'drying oils' have been recognised. All are triglycerides corresponding to (41) where OOCR represents the residue of a mixture of long chain fatty acids, some at least containing the allylic unsaturation required for curing. A methylene group flanked by two olefinic bonds is very much more effective than simple allylic site, whilst conjugated unsaturation also exerts a powerful activating effect. Some important fatty acids and their sources are given in Table 7.10.

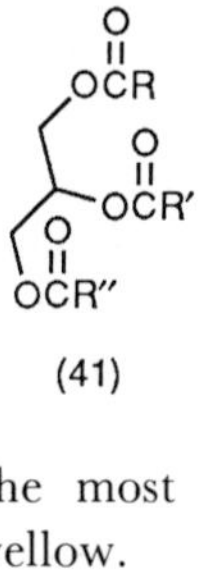

(41)

The triene-acids produce the most rapid cure, but are usually associated with a tendency to yellow.

Table 7.10 *Some common fatty acids and their sources*

Name	*Source*	*%	*Systematic name*
Stearic	—	—	Octadecanoic acid
Oleic	Palm oil	50	Octadeca-9-enoic acid
Ricinoleic	Castor oil	87	12-Hydroxyoctadeca-9-enoic acid
Linoleic	Soya oil	54	Octadeca-9,12-dienoic acid
Linolenic	Linseed oil	50	Octadeca-9,12,15-trienoic acid
Eleostearic	Tung oil	80	Octadeca-9,11,13-trienoic acid
Licanic	Oiticica oil	78	4-Ketooctadeca-9,11,13-trienoic acid

*% refers to a typical proportion of the named acid in the total

Suitable natural oils are mobile liquids which can crosslink to satisfactory coatings, however, heat treatments known as 'bodying' were developed at an early stage to improve their film forming character and film properties. The oil was held at a high temperature in contact with air, and possibly catalysts, until the required viscosity was achieved. Optionally, various natural resins were incorporated to promote hardness. The best known natural resin is 'rosin', obtained from pine-wood and consisting largely of abietic acid (42) and its isomers. 'Oleoresinous' media find only limited modern application.

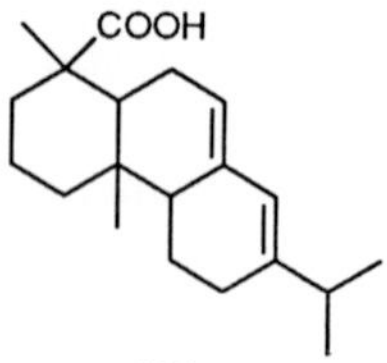

COOH

(42)

Auto-oxidative cure chemistry advanced considerably in the 1920s with the invention of 'alkyd' resins. An alkyd is a polyester with fatty acid residues attached along its backbone. Though saturated alkyds are known, and are cured through terminal acid or hydroxyl functionality, the term is usually understood to refer to polyesters furnished with auto-oxidisable side chains and providing much higher functionality than any oil, bodied or otherwise.

In modern practice, a variety of diacids, polyols, and fatty acids are reacted, and viscosity monitored, until the required product is obtained. The traditional method involved preparation of a monoglyceride from the chosen oil and additional glycerol, completion being assessed as miscibility with methanol. The monoglyceride is then processed with phthalic anhydride and optionally more oil or glycerol. The product typically contained the residue (43).

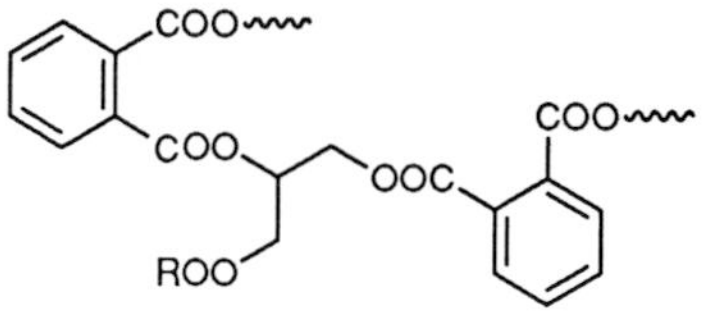

(43)

Alkyds are described as long, medium, or short oil according to the ratio of phthalic anhydride to triglyceride. Increasing oil length leads to increasing solubility in aliphatic hydrocarbons and easier manufacture. Very short oil alkyds are liable to undergo exponential increases in viscosity during manufacture.

The crosslinking rate of alkyds is markedly affected by transition metals and it is usual to add oil-soluble metal compounds such as cobalt naphthenate as 'driers' or 'siccatives'. Surprisingly, a number of non-transition metals, such as calcium, also appear to affect drying.

Unfortunately, the auto-oxidation process which leads to curing is identical with that which leads to degradation, and does not stop when the required level of cure is obtained. All alkyds tend to embrittle through their service lifetime, more so when high levels of drier have been used. They are, nevertheless, extensively used in architectural and heavy duty markets.

Other attempts to incorporate fatty acids in film formers include epoxy esters, made from bisphenol A or epoxy novolac polymers, urethane alkyds and oils made by modification of hydroxyl functional precursors with polyisocyanates, and acrylic alkyds, usually made by treating glycidyl functional acrylics with fatty acids. The products usually show property enhancements characteristic of their modifier—adhesion, toughness, and weatherability respectively.

Attempts to synthesise auto-oxidisable species have yet to provide film formers with the combination of performance, convenience, and economy which derivatives of natural fatty acids can provide. None has become widely used except as a reactive diluent (*q.v.*).

Other Homolytic Step Growth Processes. The free radical addition of thiols to olefins occurs readily on ultraviolet irradiation, or on contact with chemically generated free radicals (Scheme 56). The odour associated with thiols has prevented thiol-ene chemistry from becoming as widely exploited in coatings as might otherwise be expected.

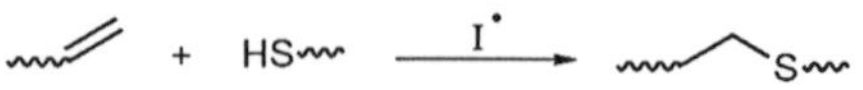

Scheme 56

A recently disclosed proprietary technology substitutes amine activated methylene for thiol (Scheme 57).

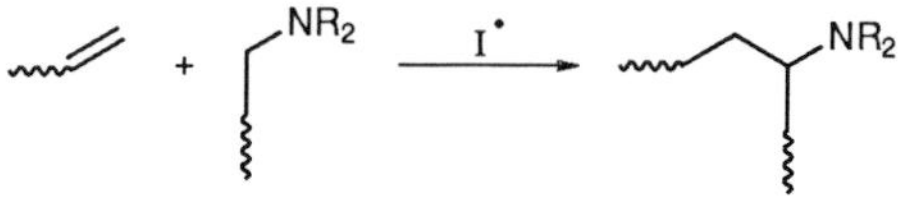

Scheme 57

Hydrosilylation reactions have also been used in crosslinking chemistry, most frequently when at least one component is a silicone polymer (Scheme 58).

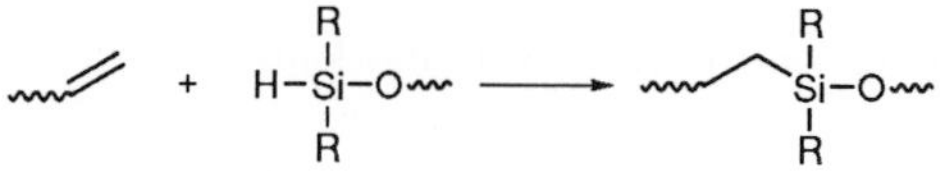

Scheme 58

Chain Growth Homolyses

Free Radical Interactions of Activated Olefins. If molecules containing two or more acryloyl groups are induced to undergo chain growth polymerisation, a crosslinked network will rapidly develop. In practice, blends of mono- and di- (or higher) acryloyl materials are formulated to provide the required network density.

Acrylate monomers are blended with diacrylates, such as hexanediol diacrylate, trimethylolpropane diacrylate or bisacryloyl derivatives of polyether, bisphenol A epoxies, or polyurethane polymers. (The nature of the crosslinking chain naturally has a profound effect on the film properties of the network.) Methacrylates generally show an unsatisfactory reaction rate.

The composition can be cured by treatment with peroxide or a low energy electron beam, but most usually are irradiated with ultraviolet light from a mercury vapour lamp. For UV cure, it is necessary to include a photo-initiator combination such as benzophenone and an amine

(Scheme 59). Alternatively, a benzoin ether can be used (Scheme 60). The radicals so produced initiate polymerisation.

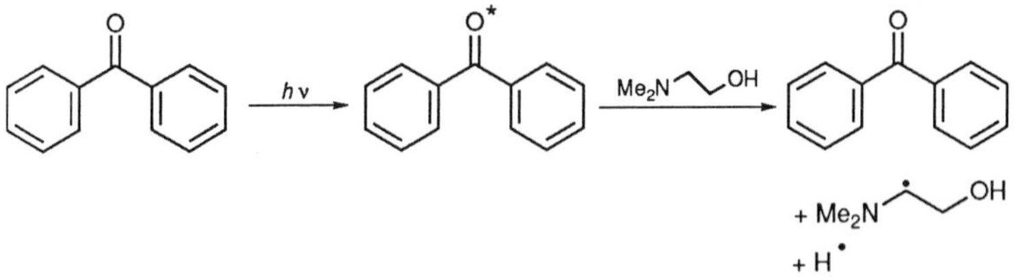

Scheme 59

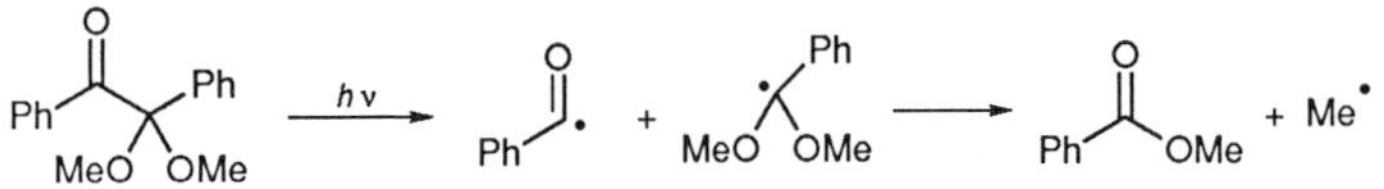

Scheme 60

UV curing has the advantage of using materials which are converted to network with no loss of volatiles. It is also very rapid and energy efficient. On the other hand, most of the available acrylate monomers are highly toxic, and the need for thorough irradiation makes it difficult to coat complex shapes, or to use very thick or pigmented layers. The inhibiting effect of oxygen at the coating/air interface also leads to stickiness which is difficult to control.

The weathering properties of UV cured systems tend to be disappointing (for polyacrylates). Both the continuing presence of photo-initiator, and occluded radical species have been suggested to explain the poor photo-stability. Nevertheless, radiation curing systems are enjoying a period of rapid growth.

Another 100% non-volatiles technology based on essentially the same principle involves the use of unsaturated polyesters. Nominally, linear polyesters are prepared from the usual diols and diacids, but including a proportion of maleic anhydride whose residues largely isomerise to fumarate during processing. The polymer is dissolved in styrene at a level of around 50%.

On adding a soluble transition metal compound such as cobalt naphthenate and a peroxide such as di-t-butyl peroxide, the styrene polymerises, alternating with fumarate residues when it can, to produce a network with polystyrene and polyester character.

Unsaturated polyesters like acrylate ester systems tend to suffer from oxygen inhibition, and their surface remains sticky after curing. The effect can be minimised in high gloss wood finishes by adding small quantities of wax which float to the surface as an oxygen barrier, and are subsequently removed by polishing. On the other hand it is turned to advantage in glass reinforced polyester technology where the usual painting process is inverted: pigmented 'gel coat' is applied to a mould in a thick layer. Cure of its surface is inhibited so that it bonds effectively with the structural polyester and glass fibre which is 'layed up' over it. On removal from the mould the former mould/polymer interface becomes the surface, free of inhibition and faithfully reflecting the smoothness of the mould.

Unsaturated polyester/glass fibre technology is widely used for the manufacture of small boats and similar objects, and a pigmented coating of similar composition is an integral part of such artefacts.

Miscellaneous Crosslinking Chemistries

Other Ring Interactions. A number of organic reactions are difficult to fit into the above classification.

Ring–ring interactions of a step growth character include several proposals to exploit the Diels–Alder 1,4-cycloadditions (*e.g.* Scheme 61). Similarly, the interaction of cyclic anhydrides with oxazolines (Scheme 62) has been suggested as a crosslinking technology.

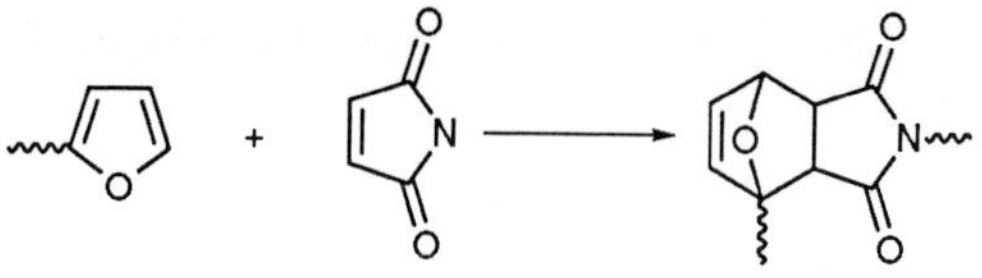

Scheme 61

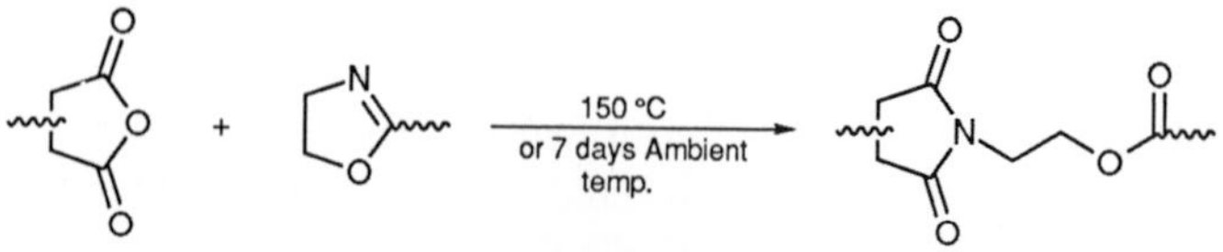

150 °C
or 7 days Ambient temp.

Scheme 62

Metal Crosslinking. The inclusion of a divalent metal salt in a polymer containing carboxylic or sulfonic acid or chelating groups such as 1,3-dicarbonyl, is widespread in elastomer and engineering plastics technology.

Such additions can benefit materials properties to an even greater extent than simple crosslinking, since ionic 'domains' radically alter the polymer morphology. Such materials are known as 'ionomers' and the crosslinks are in many cases thermoreversible, so that the product shows both thermoplastic and thermoset properties.

The use of ionic crosslinking in carboxyl functional coatings is increasing, notably in water-based systems. Preferred metals are zinc, often introduced as its oxide or acetate, and zirconium as zirconium propionate in organic systems and ammonium zirconium carbonate in water-based materials. Strongly coloured transition metals are usually avoided for obvious reasons.

Titanium and aluminium alkoxides are also popular. One proprietary system uses an acetoacetyl functional polymer in alcoholic solvent, mixed with tetrabutyl titanate. It remains tractable until the alcohol is evaporated, when titanium chelation as (44) yields a stable crosslink.

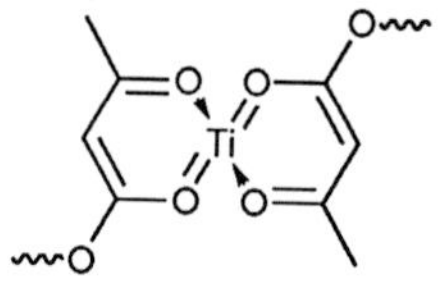

(44)

An unusual curing mechanism is used with 'etch' or 'wash' primers for the protection of structural steel in storage. It involves interactions between residual hydroxyl groups on a polyvinylbutyral polymer and zinc tetroxychromate pigment.

Dual Mechanism Crosslinking. The time-honoured practice of combining different curing mechanisms in the hope of achieving synergistic or at least additive effects is still popular. One sees, for example, proposals to include cyclic anhydride and cyclic carbonate in one molecule which is cured with an amine, or cyclic anhydride and isocyanate reacting with a polyol. Again, silanol condensation processes may be combined with a wide variety of other chemistries such as epoxy polymerisation. More significantly, combinations of thermal and radiation curing processes have potential for very marked synergy.

HIGH SOLIDS COATINGS

We have seen that reactive systems have long been used to increase the non-volatile content of coating compositions to a technically and economically convenient level, variable according to end use, but typically about 50%.

More recently, environmental concerns and legislation have demanded increasingly higher 'solids' materials, of the order of 70–80%, with a view to the eventual elimination of volatile organic solvents (*c.f.* Chapter 2).

Whilst adjustments of solvents and pigments can give significant improvements in some formulations, it is primarily the binder system which must be modified if viscosity is to be maintained as solvent levels are lowered.

Polymer Architecture

Reductions in polymer viscosity can be obtained by reducing molecular weight; by reducing the width of molecular weight distribution (since high molecular weight fractions have a disproportionate effect on overall viscosity); by changing the chemical composition of the polymer so as to reduce its glass temperature; or by reducing or removing polymer–polymer interactions, particularly hydrogen bonding. No change is without cost however. Reducing film-former molecular weight necessitates increasing levels of crosslinker; narrower molecular weight materials often give less flexible thermosets than their polydisperse counterparts; reduced glass-temperature polymers are by definition softer; and polar groupings are frequently associated with crosslinking. The impact of such changes tends to be different for different polymer types.

End-functional polymers such as polyesters experience a decrease in equivalent weight as molecular weight is reduced (since the functionality remains constant). If (45) represents a conventional polyester and (46) is its high solids counterpart, whose molecular weight may be as low as 1000, limited only by polymer volatility, it is clear that the high solids system involves less 'backbone' for each crosslink, so is likely to provide a more rigid network, offsetting the effect of any reduction in T_g.

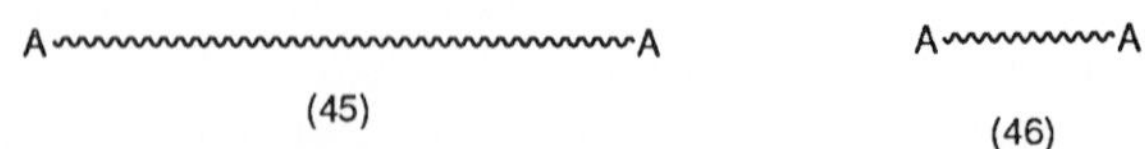

Branched polyesters tend to have wide polydispersity hence high viscosity. It follows that the quest for high solids polyesters should focus on well defined polymers with a functionality of exactly two, rendering branching unnecessary.

Higher functionalities may also be of interest and 'star' polymers, with the correct architecture, may show low viscosity for a given molecular weight since chain entanglements may be reduced in them. It is difficult to adapt a step growth process to yield well defined polymers, and chain growth procedures such as the polymerisation of caprolactone (Scheme 9) or the alternating interaction of monoepoxide and cyclic anhydride

(Scheme 63) may be considered. Unfortunately, the crystallinity of the former and the inherent hydrolytic susceptibility of the latter limit their practical use.

Scheme 63

Where laterally functional molecules are concerned, a reduction in chain length at a given composition leads to an automatic reduction in functionality [compare the conventional acrylic (47) with high solids acrylic (48)]. A less densely crosslinked network and softer coating will result, moreover, the fraction of molecules containing one functional group, or none at all, will increase sharply with even more degradation of film properties.

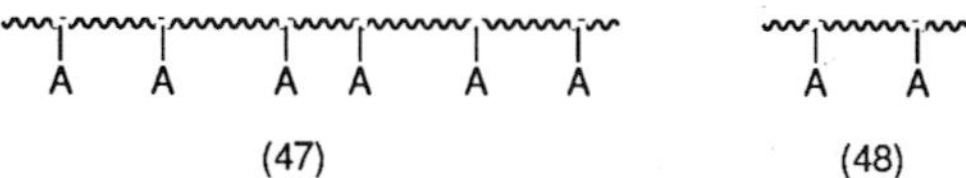

(47) (48)

Efforts have been directed towards production of acrylic polymers with at least some end-functionality. Free radical initiators or chain transfer agents (or preferably both), containing functional groups will cause the functional group to be attached to at least one end of most polymers. To the extent that chains terminate by combination, telechelic molecules can be obtained. A limited number of chain transfer agents provide functional moieties for chain termination and initiation hence reliably difunctional species. Some examples are given in Table 7.11.

More reliable specific syntheses involve ionic or group transfer techniques which bring their own practical difficulties.

Crosslinking in High Solids Coatings

As solids are increased, crosslinking chemistry plays a more important part in film formation since lacquer drying becomes less significant. In the limit, a coating network may be constructed entirely from low molecular weight oligomers with no intrinsic film-forming character. It follows that features of the reactive groups and new linkages which are regarded as desirable in conventional systems are now imperative—they must be

Table 7.11 *Functional initiators and chain transfer agents*

	Functional terminations
INITIATORS	None
N=N CN CN	
HOOC NC N=N CN COOH	—COOH
Et CN HO N=N OH NC Et	—OH
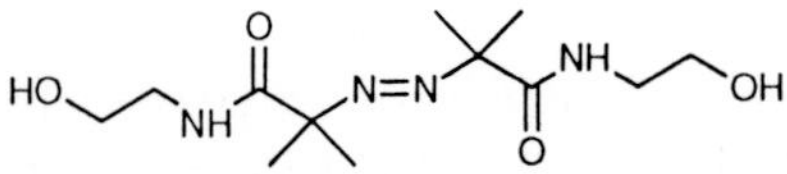 O HO NH N=N NH OH O	—OH
CHAIN TRANFER AGENTS	None
$C_{12}H_{25}SH$	
HOOC SH	—COOH
HOOC SH	—COOH
HOOC SH	—COOH
HO SH	—OH
S S HO O SS O OH*	—OH
HO S S OH* N SS N HO OH	OH —N OH

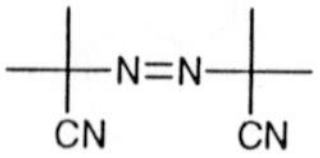

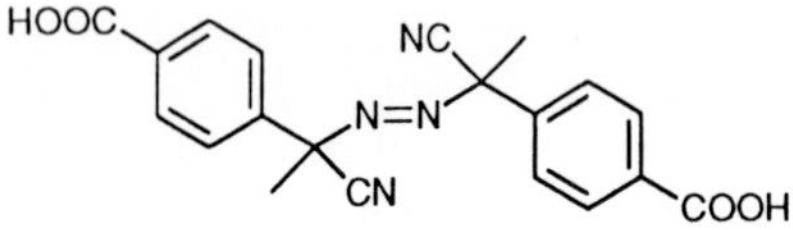

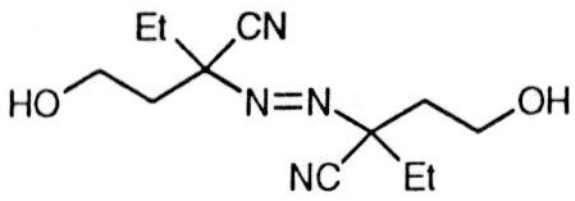

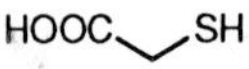

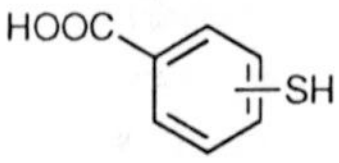

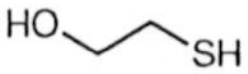

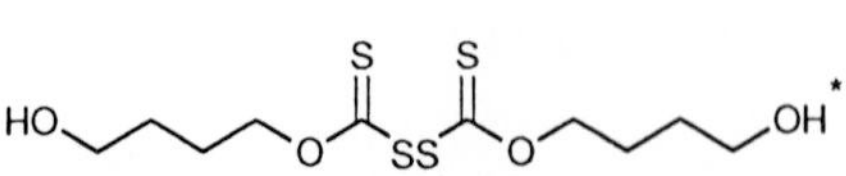

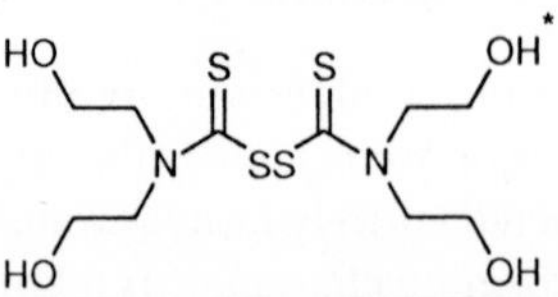

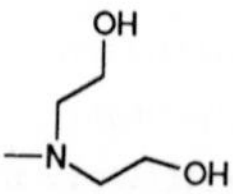

*Disulfides provide identical fragments for chain termination and initiation hence a greater likelihood of functionalising both ends.

stable, react at the correct rate, possess acceptable toxicology, and make the required contribution to the overall network character.

The relationship between the reaction rate in the liquid condition and the drying film (pot life: cure time) is altered since the high solids paint will necessarily contain a higher concentration of functional groups which will react at a correspondingly higher rate.

It is also necessary to consider secondary interactions between reactive groups. The familiar functions such as hydroxyl and carboxyl with their H-bonding potential may give rise to unacceptable levels of polymer–polymer interaction. Nevertheless, isocyanate and amino-formaldehyde cured polyol chemistry is widely practised in successful high solids compositions. In the latter, condensed alcohol makes a significant contribution to organic volatiles, and it is usually desirable to use methylated aminos to minimise its mass, though butylated aminos may be less viscous. Higher solids are also obtained from the better characterised reactions of fully methylated systems with specific acid catalysts. Glycoluril derivatives are particularly promoted since they can only undergo specific acid catalysed reactions.

One innovation which is gaining acceptance is the use of active-methylene cure chemistries. In addition to various Michael reaction possibilities, acetoacetyl groups can react with amino-formaldehyde, isocyanates, amines, and aldehydes (Scheme 64). Since they exist as

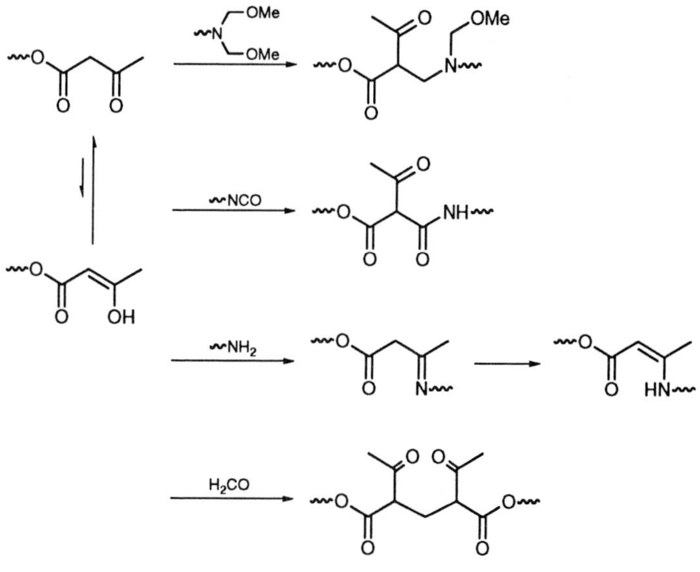

Scheme 64

keto–enol tautomers with the enol very much in the minority, their H-bonding potential is very much lower than that of hydroxyl groups. Acetoacetyl functionality can be introduced to polymers containing hydroxyl groups by treatment with acetoacetate esters or diketene.

Reactive Diluents

A reactive diluent is a molecule with solvent like properties which ultimately becomes part of the thermoset network, so does not contribute to volatiles.

In a reactive system containing A and B functional groups for crosslinking, RA could be an effective reactive diluent, but would consume B groups, upsetting the stoicheiometry. For example, in an isocyanate/polyol system, addition of butanol would consume expensive isocyanate without contributing to network structure and it might even prevent network formation.

It is more satisfactory to supplement or even supplant the crosslinker with a difunctional diluent, A–A.

More attractive still are molecules of the type A–B which can form bridges between A and B polymeric groups without disturbing the overall stoicheiometry. The monofunctional acrylate and styrene used in radiation cured systems and unsaturated polyesters respectively represent the AB principle in highly successful 100% non-volatile systems.

Taking into account the possibility of using reactive diluents independent of the cure chemistry, or in otherwise non-convertible systems, reactive diluents can be classified as in Table 7.12.

Reactive diluents in stoved systems require low volatility and high reactivity, since vaporisation represents failure, and the reactive materials used may be even less environmentally desirable than conventional solvents. Polyols, polycaprolactone–polyols, and caprolactone itself have been used in melamine–formaldehyde cured polyol systems.

The use of low molecular weight aliphatic epoxies, popular in potting compounds for example, is generally undesirable in coatings in view of their volatility and toxicity. However, recent interest has focused on a naturally occurring epoxidised triglyceride, vernonia oil (49). Epoxy

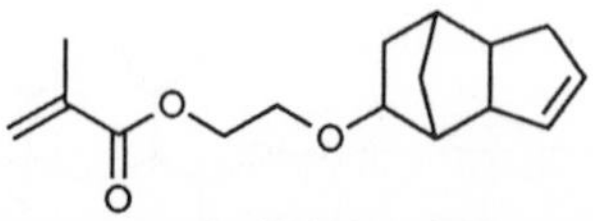

(50)

Table 7.12 *Classification of reactive diluents (A, B represent functional groups involved in crosslinking a film former)*

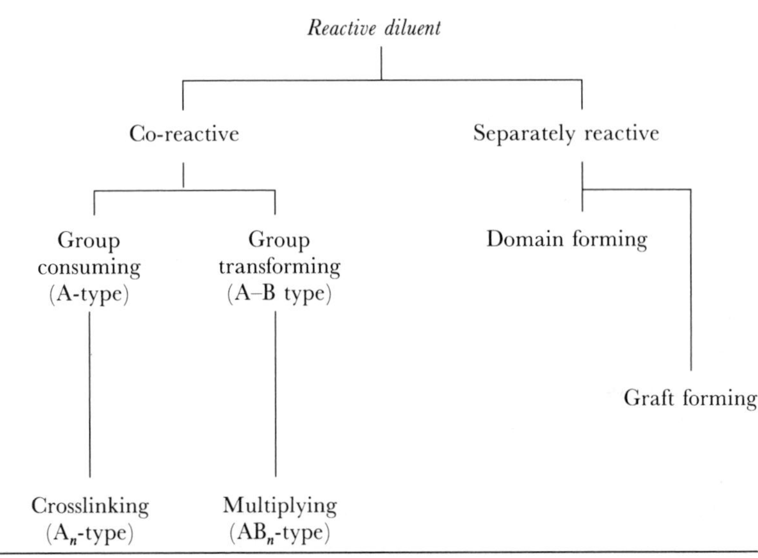

analogues have also been used, examples being polyacrylates, maleates, and cyclic carbonates which can react with amine functional curing agents.

Use of auto-oxidisable diluents with alkyds has been much more widely practised. Linseed oil itself is useful as are many other allylic esters and ethers. One popular commercial product (50) includes allylic and methacrylate groups in the same molecule. Presumably radicals produced by auto-oxidation initiate polymerisation and grafting of the methacrylate with consequent benefit to general film properties.

R = —$(CH_2)_7$—CH=CH—CH$_2$—(epoxide)—$(CH_2)_4CH_3$

(49)

Another commercial product is a low molecular weight polymer of allylglycidyl ether, a polyether with pendant allyl groups. It can be used in oxidising systems or in free radical systems such as UV cured acrylates. In the latter it is said to enhance the free radical flux to the point where oxygen inhibition is not apparent.

POWDER COATINGS

As with the corresponding liquid paint systems, thermosetting powder coatings can show material advantages over their thermoplastic counterparts since lower melt viscosities can be achieved, with consequent reduction in film thickness and improved appearance. Thermoplastic powder coatings are generally of high molecular weight polymers such as polyethylene or polyamide and have rather restricted application.

Polymers for Powder Coatings

Thermosetting powders are based on more familiar coatings polymers—epoxies, polyesters, and, occasionally, acrylics formulated to have a glass temperature above 40 °C to ensure powder stability. Otherwise, the design considerations affecting high solids paints are applied; polymers of relatively low and well defined molecular weight are used, and crosslinking kinetics requires careful attention.

Since powder coatings are normally manufactured in heated extruders, chemistry and rheology have to conform to stringent constraints illustrated in Table 7.13 (*c.f.* Chapter 4). The need to be chemically stable at extrusion temperature ensures that stoving temperature can only be reduced below 150 °C with great difficulty.

The earliest powder coatings were based on solid bisphenol A epoxy resins, and they are still widely used where exterior weatherability is not important. The largest part of the exterior market however is served by polyesters based on terephthalic or isophthalic acids and neopentylglycol as preferred diol (*c.f.* Table 7.1) with low levels of branching. 'Hybrid' systems containing comparable levels of co-reactive polyesters and epoxy resins provide intermediate properties.

Powder Crosslinking

Complete absence of volatiles is a key requirement for powder coatings since the presence of any organic vapour in a stoving oven would necessitate increased ventilation with consequent loss of energy efficiency. It follows that ring opening crosslinking reactions are preferred.

The earliest crosslinkers for epoxy resins were dicyandiamide (51) and its derivatives. Their chemistry is complex, but all four active hydrogens on (51) can participate, cyclisation of the products is known to occur, and an activated nitrile/hydroxyl reaction has also been detected. Epoxy self-additions are likely to be promoted by the strongly basic amidine.

Table 7.13 *Thermal constraints affecting powder coatings*

Temp(°C)	OPERATION	PHYSICAL	CHEMICAL
	Stoving	flow and coalesce	crosslink
150			
100	Extrusion	be plastic	be stable
50			
	Storage	be stable	be stable

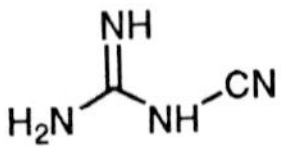

NH

H_2N NH CN

(51)

Carboxylic anhydrides and acids have achieved considerable popularity, materials such as phthalic anhydride, trimellitic anhydride (Table 7.1) or its esters, such as (52) and pyromellitic (53) and benzophenonetetracarboxylic acid (54) bis anhydrides giving well cured networks with epoxy resins. The trimellitic esters have the advantage that melting points and tendencies to produce harmful dust, are reduced.

Acid tipped polyesters are cured with epoxies, most notably the weatherable triglycidylisocyanurate (TGIC, 55), catalysts being selected from a wide range of amines, amidines, and phosphines. TGIC is the staple curing agent for weatherable powder coatings at the time of writing, but concerns about its toxicology are causing the Industry to look for replacements. When epoxies derived from bisphenol A are used with acid tipped polyesters, 'hybrid' coatings of intermediate weatherability are obtained.

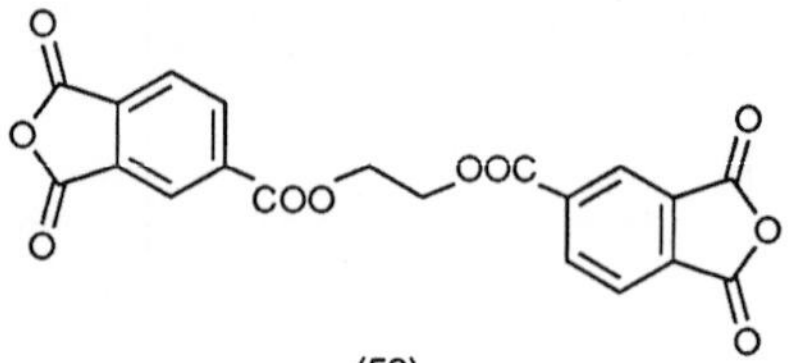

COO OOC

(52)

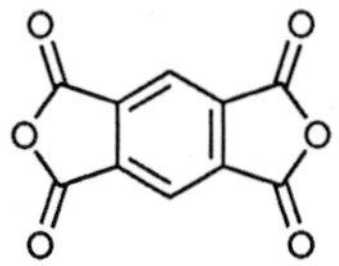

(53)

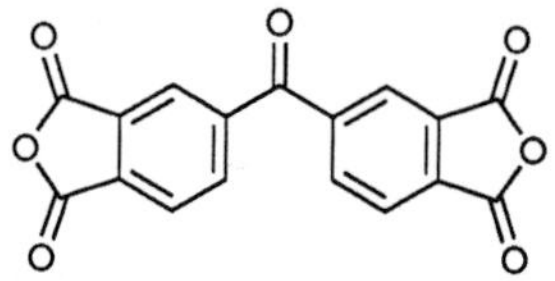

(54)

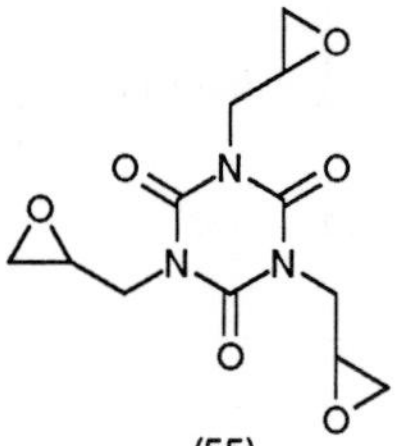

(55)

Blocked isocyanates have enjoyed some importance in powder coatings, undesirable volatiles notwithstanding, and hydroxyl tipped polyesters for example can be cured with caprolactam blocked adducts of isophoronediisocyanate (see Table 4.8). The resulting films often show better flow and levelling than comparable systems, perhaps because of the kinetics of the reaction, or the fluxing effect of the liberated caprolactam. The latent isocyanates, which do not generate a volatile by-product, naturally attract particular interest in powder coatings circles and uretidinediones (see Scheme 17) have been claimed as particularly useful crosslinkers.

Other condensation systems recently advanced as candidates include β-hydroxyamides, for example (56), which prove to react with acid groups much more readily than non-activated polyols, and glycoluril/formaldehyde condensates.

(56)

CONCLUSION

Polymers and their manipulation are at the very heart of coatings. They can be adapted to give hard, elastic, resistant, or removable systems of almost unlimited variety. The thrust of much current development work is towards polymers for use in coatings which provide the required performance with the minimum impact on the environment.

BIBLIOGRAPHY

J. W. Nicholson, 'The Chemistry of Polymers', Royal Society of Chemistry Paperbacks, The Royal Society of Chemistry, Cambridge, 1991.

'Resins for Surface Coatings', eds. P. Oldring and G. Hayward, SITA Technology, London, 1987.

S. Paul, in 'Comprehensive Polymer Science', eds. G. Allen and J. C. Bevington, Pergamon Press, Oxford, 1988, p. 149.

Z. W. Wicks Jnr. and L. W. Hill, 'Design Considerations for High Solids Coatings', *Prog. Org. Coatings*, 1982, **10**, 55.

T. A. Misev, 'Chemistry and Technology of Powder Coatings', John Wiley and Sons, Chichester, 1991.

Chapter 8

Disperse Phase Polymers

W. A. E. DUNK

INTRODUCTION

A fully formulated paint is a complex system which comprises pigments, solvents, and a variety of additives. All are dispersed in a continuous vehicle that may be an aqueous or non-aqueous solution or colloidal dispersion of a polymer that constitutes the binder. These binders largely determine the character of the coating, and it should be noted that in the present time of increasing attention to atmospheric pollution, the use of solvents and solvent based binders is not encouraged (*c.f.* Chapter 2). The use of water-borne binder resins is expanding and this chapter will emphasise these systems.

Before considering the dispersions, a word on the colloidal state is in order. A colloid may be defined as any system that contains a physical boundary dimension in the range 1 nm to 1000 nm. Such a system may be differentiated from suspensions in a number of ways but for simplicity it is sufficient to remember that a colloidal dispersion is not influenced by gravity whereas a suspension is. It will also be helpful to consider the differentiation of colloids into those that are solvent loving (lyophilic) and those that are solvent hating (lyophobic); when water is the continuous phase the terms hydrophilic and hydrophobic are used. In the case of a solvent or water soluble polymer, the result is a lyophilic colloid system, whereas when the polymer exists as finely divided solid matter dispersed in an organic solvent or water, we have a lyophobic colloid system.

In the coatings industry there is a degree of ambiguity and obscurity in the definition of colloidal dispersions, and this is most commonly seen in the use of the terms emulsion, latex, dispersion, and suspension which are often used synonymously. Table 8.1 defines these terms as they are understood here.

Table 8.1 *Definitions of terms encountered in disperse phase polymer science*

Dispersion	A distribution of finely divided solid particles in a liquid phase to give a system of very high solid–liquid interfacial area.
Emulsion	A heterogeneous system consisting of at least one immiscible liquid dispersed in the form of droplets in another liquid.
Latex	A colloidal dispersion of a solid polymer in a liquid continuous phase. Such a system should be termed a latex only when it has been prepared by emulsion polymerisation.
Suspension	A non-colloidal dispersion of a solid in a liquid. In such systems the solid–liquid interfacial area is low and size is $>1\ \mu m$.

It is useful to recall that latexes as defined in Table 8.1 are all dispersions, but all dispersions are not latexes. Similarly, the definition of emulsion is sufficient to preclude its use in describing latexes as emulsion polymers, a common mis-use frequently seen in trade literature. Presumably this comes about because the process by which they are made begins with the emulsification of liquid monomers which are then polymerised.

In this chapter we shall consider a number of dispersion systems that have been used in the coatings industry, but the emphasis will be directed towards those that are water-based, with particular attention to emulsion polymerisation.

A brief introduction will also be made to the technique of suspension polymerisation since this may have application for the synthesis of resins for powder coatings. However, our main concern will be with the enormously important industrial process of emulsion polymerisation.

Before embarking on a descriptive tour of the systems presented, it is as well to draw attention to the advantage of dispersion systems over solutions in terms of viscosity–molecular weight relationships. It has been shown, in Chapter 5, that film strength increases as molecular weight increases, hence high molecular weights are generally desirable. In solution, such high molecular weight polymers would need to be present at uneconomically low concentrations to attain acceptable application viscosities. This is reflected somewhat in the water reducible systems though, due to the pseudo-dispersion nature of these types, molecular weights of 20 000–50 000 may be used at acceptable concentration levels.

At the other extreme, latexes have their disperse phase polymers with molecular weights commonly over 1 000 000 but, in this case, viscosity is independent of molecular weight. The factors that do affect viscosity in such systems are polymer particle shape and the packing factor which, along with the volume fraction of the dispersed polymer, can be expressed in mathematical form, the best known of which is the Mooney equation:

$$\ln \eta_r = \frac{k\phi}{(1 - s)\phi} \tag{1}$$

where η_r is the relative viscosity, ϕ is the volume fraction of disperse phase, s is the self crowding factor, and k is the Einstein coefficient, 2.5.

Other colloidal dispersions are somewhere in between the water reducibles and latexes with respect to viscosity–molecular weight dependence.

EMULSION POLYMERISATION

The importance of latexes in many industrial applications, and particularly in the coatings industry where water-borne systems have increasing importance, warrants a somewhat more extensive coverage than the other disperse phase systems that have been given here.

In 1982, George Ham commented that there had been more studies made in the field of emulsion polymerisation than in any other area of polymer chemistry and this is probably as true ten years later as it was then. There can be little doubt that the industrial importance of the procedure plays a large part in this, and this can be attributed to the heterogeneity of the reaction system that allows high molecular weight polymers to be synthesised at an economical rate of reaction, a situation not available in bulk, suspension, or solution polymerisation. It is the purpose of this section to review these aspects at an introductory level, with emphasis on oil in water emulsion systems since the water in oil (or inverse) technique has little application to coatings.

After a brief historical introduction, we shall consider qualitative and quantitative theories that have been proposed, then present some practical aspects of the process. A short consideration of latex properties will be given and we shall close with a review of more recent developments, particularly the consequences of sequential polymerisation. At this point it is worth recalling the comment of Vanderhoff that, although the practice of emulsion polymerisation can be based on kinetic principles of free radical initiated addition polymerisation, the complex interactions of monomers, initiators, emulsifiers and other ingredients leave the quality of the final latex very much in the hands of the operator.

Historical Background

The first approach to the preparation of a latex may be traced back to the work of Hofmann and his collaborators at Bayer in the first decade of this century. His objective was to imitate the physiological conditions under which rubber is formed in the tree and utilise those conditions to prepare synthetic rubber from isoprene or butadiene. These monomers were mixed with water in the presence of albumin or gelatine as protective colloids and heated over periods of days and even weeks. Polymerisation occurred, probably through peroxides randomly formed from the monomers and atmospheric oxygen. This work was disclosed in patents but few studies were published in this area until after World War I.

Somewhat unsystematic studies were made in the early 1920s when it was observed that soaps such as oleates, alkylarylsulfonates, and abietates could be used to form stable monomer emulsions that led to stable latexes. Paralleling this work, Luther showed that by introducing certain compounds such as hydrogen peroxide, persulfates, and perborates, the long polymerisation times could be reduced to hours. Probably the first real emulsion polymerisation to be based on a rational recipe was that of Dinsmore who disclosed the preparation of poly(dimethylbutadiene) in 1929. Subsequently the use of this technique grew enormously, urged on by the need for synthetic rubber, although the first technical application appears to have been for an acrylic latex for leather finishing (I. G. Farben, 1931). Other latexes were produced for adhesives, coatings, synthetic leather, and many other end uses. Table 8.2 gives some examples of polymers prepared by this technique.

Table 8.2 *Some polymers prepared by emulsion polymerisation*

Monomer	*Year*	*Patent reference*
Acrylic Esters	1930	Ger 654 989
Vinyl Ether	1930	Ger 634 408
Vinyl Chloride	1932	US 2068 424
Vinyl Esters	1934	Ger 727 955
Ethylene	1938	Ger 737 960
Vinylidene Chloride	1951	*Makromol Chem*, 1951, **6**, 39

The interested reader would find an article in *Industrial and Engineering Chemistry*, 1933 (G. Whitley *et al.*, **25**, 1204, 1388) fascinating reading, not least for the variety of materials—milk, glue, and even blood—which were included in emulsion polymerisation recipes.

The complexity of a typical emulsion polymerisation recipe has been mentioned earlier in this review and the formulation of an industrial procedure can be a formidable task. In addition to the continuous water phase, monomer(s), initiator, and emulsifier there may be a buffer and chain transfer agent and even a second initiator and surfactant. The efforts extended over the years to understand even the simplest systems, and the level of current understanding of them, make Vanderhoff's comment on the role of the operator even more vivid. We shall now give an introduction to qualitative and quantitative theories before moving to practical aspects.

Qualitative Theories

The mechanism of emulsion polymerisation has been particularly concerned with the locus of particle nucleation and the growth of polymer particles. In 1935 Staudinger proposed that the locus of nucleation was at the surface of the monomer droplets but this was subsequently shown to be erroneous. A few years later Fikenscher believed that polymerisation commenced in the aqueous phase, monomer dissolved in the water being initiated by the free radicals generated in that phase and the monomer concentration was maintained by dissolution of more monomer from emulsified monomer droplets. In the presence of a soap, monomer would be solubilised in micelles and, even as long ago as 1938, these monomer swollen micelles were believed to be the principal locus of polymerisation.

During World War II several researchers independently published qualitative pictures of the emulsion polymerisation process, but it is Harkins to whom the classical picture is attributed. The proposed picture was one of a compartmentalised system of monomer(s), water, emulsifier, and initiator, where the monomer has only slight water solubility and the free radical generator is in the continuous phase, *i.e.* a water soluble initiator. Figure 8.1 gives a picture of such a system and it should be noted that the monomer(s) may be in 3 domains:

(i) dissolved in the continuous water phase;
(ii) contained in surfactant micelles; or
(iii) as emulsified monomer droplets.

Similarly, the surfactant may be found at each of these domains, but once initiation of polymerisation has commenced, the emulsifier will be found at the surface of monomer swollen polymer particles.

Conceptually similar to initiation in micelles is the Medvedev mechanism of initiation in the adsorbed emulsifier layer. This may be true for some systems but, as Napper has pointed out, it is not possible to differentiate the mechanisms by kinetic studies.

Once free radicals have been generated, either by thermal decomposi-

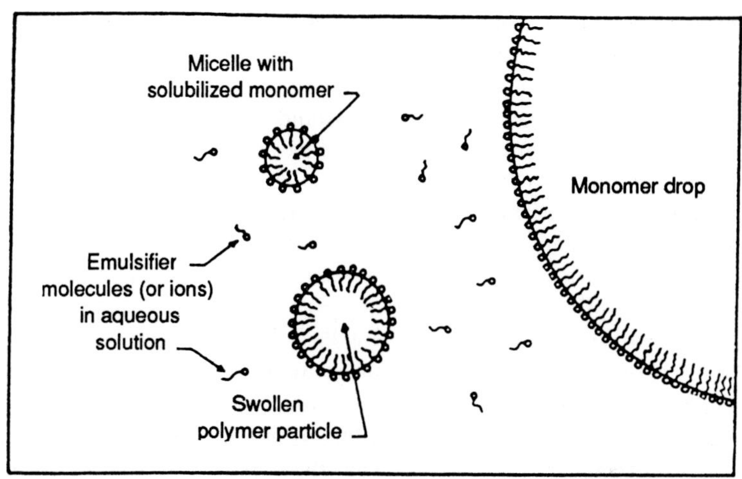

Figure 8.1 *The Harkins Picture of emulsion polymerisation (water soluble initiator and dissolved monomer not shown)*
(From J. W. Vanderhoff *et al., Adv. Chem.*, 1962, **34**, 32)

tion of the initiator or by addition of an activator in the case of redox couples, initiation of the monomer(s) may occur by two distinct mechanisms:

(i) micellar nucleation where a free-radical enters a monomer swollen micelle and creates a monomer radical or
(ii) homogeneous nucleation where free radicals generate monomer radicals from monomer dissolved in the water.

In the second case the oligomeric polymer radicals may either enter monomer swollen micelles or they may propagate to a point where they reach water insolubility and precipitate, further growing by monomer adsorption, to become primary particles stabilised by emulsifier adsorption. That this mechanism is viable may be demonstrated by the observation that very insoluble water soluble monomers, *e.g. p*-t-butyl styrene, stearyl methacrylate, and most fluorinated monomers are not polymerised under standard emulsification conditions.

When free radicals enter monomer swollen micelles, the mechanistic picture is readily envisaged in terms of the difference in size of the monomer domains. Micelles consisting of typically *ca.* 100 emulsifier molecules are present in an emulsion polymerisation procedure at some 10^{18} units ml^{-1} and have a size of *ca.* 5 nm when swollen with monomer. By comparison, the stabilised monomer droplets are present in far smaller number, *ca.* 10^{10} units ml^{-1}, and have a size of *ca.* 1000 nm. Not only does the preponderance of monomer swollen micelles lead to their being the principal target for initiation, but their much larger surface area as compared to the monomer droplets enhances the capture of free-radicals.

Once polymerisation has commenced in the micelles, monomer is replenished by diffusion through the aqueous phase from the monomer droplets. Micellar swelling continues through polymer growth until the size reaches *ca.* 100 nm diameter and their number becomes *ca.* 10^{14} cm^{-3}. As the monomer droplets are replaced by polymer particles, the average particle diameter of the system is reduced, leading to an increase in total surface area. Increasingly more of the emulsifier is adsorbed until a point is reached where the micelles disappear completely and the surface tension of the system becomes constant.

At the point where the micelles disappear, the monomer conversion is between 10 and 20%, at which time the reaction rate becomes constant, *i.e.* zero order with respect to monomer. Since no new latex particles can now be generated, the monomer concentration is kept constant by the monomer droplet reservoir and when all the monomer is in the latex particles, the reaction becomes first order with respect to monomer. The stages described here and illustrated in Figure 8.2 are referred to as Intervals and give only qualitative data on emulsion polymerisation kinetics.

In leaving this view of emulsion polymerisation, we should consider these mechanisms as limiting cases and note a third theory, that of congulative nucleation advanced by Feeney *et al.* in the early 80s. This described the formation of primary particles by either of the previously described mechanisms, followed by their limited aggregation in a two step process.

Quantitative Theory

The Harkins picture of emulsion polymerisation led to attempts, notably by Haward, and Smith and Ewart, to place it on a quantitative basis, with the latter workers' results being regarded as the canonical theory of the process. Although idealised and not explaining all the observed phenomena, it is necessary to understand the ideas presented before one may appreciate later developments in theoretical interpretations. Here is given an outline of what Smith and Ewart proposed, but for a clear and comprehensive view of the theory and its later modifications, the treatise of Blackley and other references given in the Bibliography must be sought.

Smith and Ewart considered the number of particles formed in Interval I and the number of free radicals per particle present in Interval II (see Figure 8.2). Although monomer conversion is only 10–20% in Interval I,

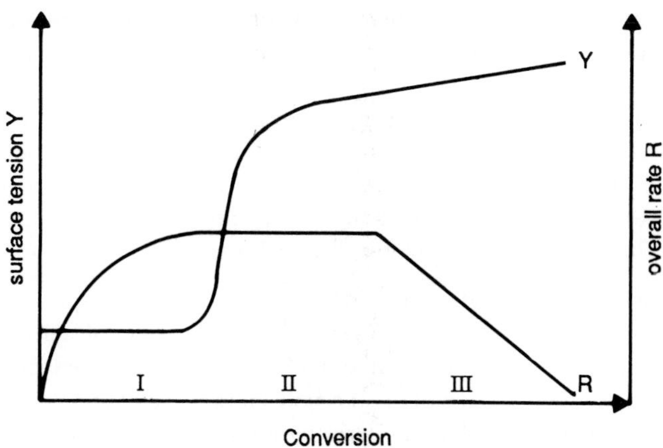

Figure 8.2 *The progress of an emulsion polymerisation in terms of surface tension, overall rate of polymerisation, and conversion*

most of the latex particles are generated here and this number may significantly influence the course of the subsequent Intervals.

In the period where zero order reaction and overall reaction rate is constant, monomer concentration, [M], particle number, N, and the number of latex particles cm^{-3} with n active radicals, N_n will be constant. When this stationary state is fulfilled, the three reactions leading to formation of latex particles with n free radicals are:

(i) entry of a radical in a latex particle with $(n - 1)$ radicals;
(ii) exit of a radical from a latex particle with $(n + 1)$ radicals; and
(iii) termination between two radicals in a particle with $(n + 2)$ radicals.

These reactions will be equal to those reactions which cause their disappearance, *i.e.* entry, exit, or termination. Those situations are symbolically expressed in the recursion equation known as the Smith–Ewart equation (2).

$$\left(\frac{C'}{N}\right) + N_{n+1}\left(\frac{K_s S}{N_A}\right)\left(\frac{[n+1]}{V_s}\right) + N_{n+2}\left(\frac{K_t}{N_A}\right)\left(\frac{[n+2][n+1]}{V_c}\right)$$

$$= N_n\left\{\left(\frac{C'}{N}\right) + \left(\frac{K_s S}{N_A}\right)\left[\frac{n}{V_s}\right] + \left(\frac{K_t}{N_A}\right)\left[\frac{n(n-1)}{V_s}\right]\right\} \qquad (2)$$

where N_A is Avagadro's number; K_s is the radical exit rate constant, mm mol^{-1} s^{-1}; S is the surface area of swollen latex particle, m^2; K_t is the radical mutual termination rate constant, $1\,mol^{-1}\,s^{-1}$; V_s is the volume of swollen polymer particle, l; C' is the rate of entry of free radicals into latex particles, $l^{-1}\,s^{-1}$.

The intractable nature of this equation led its formulators to consider just three limiting cases for which explicit solutions were obtained, although it should be noted that later Stockmeyer presented a general analytical solution, as did O'Toole a few years afterwards.

The three cases defined by Smith and Ewart were:

Case 1 where $n \ll 1$ Here the rate of exit of free radical species is very much greater than their rate of entry. This behaviour is not commonly observed but vinyl acetate is the best known example.

Case 2 where $n = 0.5$ Here the termination rate is greater than the free radical entry rate and the exit of free radical species is negligible. Styrene is the classic example of such behaviour.

Case 3 where $n \gg 1$ This situation occurs when the rate of radical entry is greater than the rate of termination. Polymers having crystalline structures at polymerisation temperature exhibit this behaviour, vinylidene chloride being the best known. It is often observed in large polymer particles.

Prediction of certain variables, *i.e.* surfactant concentration, temperature, and initiator concentration, on the reaction rate (R_p)and degree of polymerisation (DP) are possible and the theory allows the derivation of expressions for these parameters, equations (3) and (4).

$$R_p = K_p[M]\bar{n}N \tag{3}$$

$$DP = \frac{K_p[M]\bar{n}N}{R_i} \tag{4}$$

where K_p is the rate constant for propagation; N is the number of particles; $\bar{n}$ is the average number of radicals per particle; [M] is the monomer concentration; R_i is the rate of radical generation.

These kinetics are attained when the radicals are segregated and the number of loci available for segregation is within a few orders of magnitude of the number of free radicals. In contrast to the inverse variation of R_p and DP observed in bulk, suspension, and solution polymerisation, the kinetics of emulsion polymerisation allow the rate of polymerisation and degree of polymerisation to increase simultaneously, clearly a great advantage in industrial procedures.

In the forty years since the Smith–Ewart theory was published, much work on homogeneous and heterogeneous nucleation has been carried out on both practical and theoretical levels. Prediction of molecular weight distributions has also seen a considerable advance.

Practical Aspects of Emulsion Polymerisation

Having considered the theoretical aspects of emulsion polymerisation we should now devote some time to the practical aspects. Firstly let us recall the four principal components:

(i) the continuous aqueous phase;
(ii) the monomer or monomer mixture;
(iii) the free-radical source, *i.e.* the initiator; and
(iv) the surfactant or emulsifier.

Buffering agents may be needed for certain systems and control of molecular weight could necessitate the incorporation of a chain transfer agent such as a mercaptan. However, for the first trials the main components must be put together in such a way as to give a stable latex. Although later we may opt for a continuous or semi-batch procedure, it is convenient to carry out the preliminary work under batch conditions, the classic method being to seal the emulsion of monomers, along with the initiator, into a soda-pop bottle and rotate it in a water bath at the desired temperature until polymerisation is complete. Sufficient information can be derived regarding kinetics, conversion, and stability to transfer the experiments to a reactor fitted as shown in Figure 8.3. Further entry ports may be added as desired.

A typical latex recipe is likely to fall within the limits indicated here: hydrophobic monomer 40–60% volume; water 40–60% volume; water soluble initiator 0.1–1.0% by weight on monomer; emulsifier 1–5% by weight on monomer.

The steps to be taken in evolving a suitable polymerisation recipe may be considered as:

(i) Selection of a surfactant that will not only give a stable emulsion but will effectively prevent coagulation of formed polymer particles. It is important that the critical micelle concentration (c.m.c.) be known and this may be determined by measuring the surface tension of a number of solutions of surfactant at varying concentrations, the c.m.c. being that concentration where surface tension becomes constant. Very often, however, it is possible to look up this figure in, say, the Polymer Handbook or surfactants manuals. Usually one wishes to work in a micellar system and it is

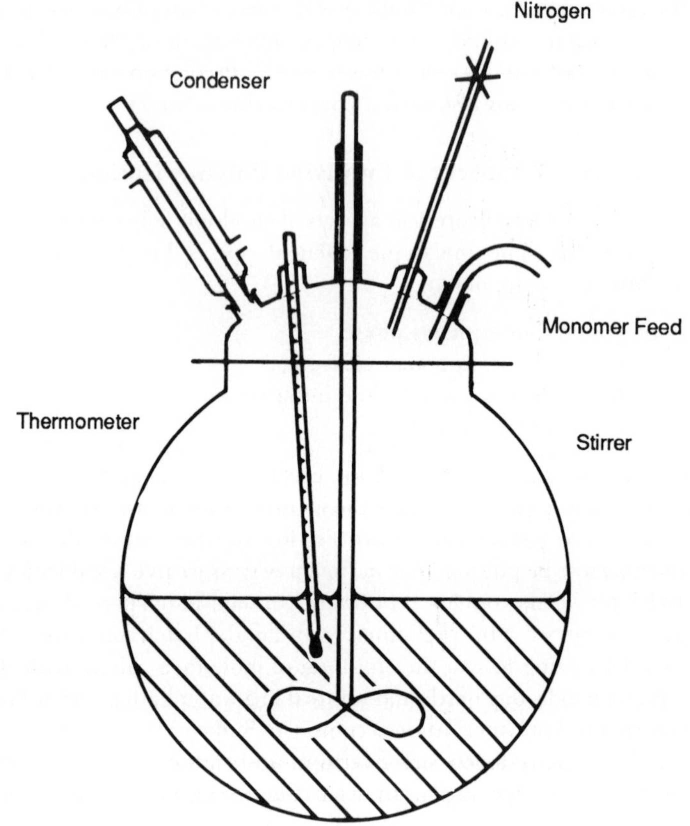

Figure 8.3 *Typical emulsion polymerisation apparatus*

best to work at the minimum concentration of surfactant that gives the desired result. Of course sub c.m.c. may be used for certain cases.

(ii) The temperature at which it is desired to conduct the polymerisation will determine the initiator selection should a thermally decomposed type be chosen. It is a useful rule of thumb to select an initiator having a half-life approximately one-third of the time needed for the polymerisation, *e.g.* for a polymerisation of 3 hours then an initiator having a half-life of one hour at the temperature desired should be chosen. When polymerisation needs to be conducted at low temperature, below $\sim 40\,^{\circ}\mathrm{C}$, redox couples are advantageously employed. The persulfate–meta bisulfite reaction is a well known example of such a couple.

(iii) Recalling that both the degree and rate of polymerisation may be increased simply by increasing the number of particles in the system, it is a useful guide to remember that, approximately, the following conditions hold:
   (a) increasing emulsifier concentration causes N to increase proportionally to [emulsifier concentration]$^{3/5}$;
   (b) overall reaction rate is proportional to N, hence to [emulsifier concentration]$^{3/5}$;
   (c) increasing initiator concentration results in N and the overall reaction rate increasing with [initiator concentration]$^{2/5}$.

By adding monomer and initiator continuously through a polymerisation reaction leads to a broad and sometimes bi-modal distribution. If a part of the monomer is emulsified and polymerised, then the remainder of the monomer added continually in the form of an emulsion, the distribution is narrowed considerably.

The number of particles nucleated is difficult to reproduce on the industrial scale but this problem can be avoided by the use of a seed latex. In this procedure, a previously prepared latex of desired particle size is used as the medium for the polymerisation of a new batch of monomer with emulsifier and initiator concentrations controlled to avoid the generation of a new crop of particles. With the polymerisation commencing at the particle growth stage, the difficultly controlled nucleation step is avoided.

Particle size and particle size distribution in latexes, and indeed in other forms of polymer dispersion, can have important effects on film properties, *e.g.* gloss and control may be affected by varying the method of polymerisation. The influence of some variables is shown in Table 8.3. However, the seeding technique is preferable, once the desired size and distribution have been obtained.

Table 8.3 *Factors affecting particle size distribution in emulsion polymerisation*

Variable	*Size distribution*
Increase emulsifier concentration	Becomes less uniform
Increase initiator concentration	Becomes more uniform
Increase monomer: water ratio	Becomes more uniform
Increase temperature	Becomes more uniform if activation energy of propagation > activation energy of initiation and *vice-versa*.

A particularly interesting advance that has important uses in the coatings industry is the formation of structured polymer particles by sequential polymerisation. It consists of polymerisation of a monomer under emulsion polymerisation conditions then adding a second monomer and polymerising it taking care that no new particles are initiated. The resulting polymer may have an essentially core–shell structure where the core may be hard and the shell soft, or the reverse.

Although latex particles are usually spherical, many other shapes are possible, and lobed polymers have been prepared by Rohm and Haas to give improved rheology in latexes for paint application. Attempts to prepare latexes with high, ≥60% polymer content, without the high viscosities normally encountered, have been made by designing polymerisations that give rise to bi-modal size distribution.

The Application of Latexes to Coatings

The advantages of latexes as binders in coatings have been noted, in particular the possibility of having high molecular weight polymers at conveniently economic concentrations, yet not paying the penalty of unmanageably high viscosities that such polymer solutions would exhibit. However the need to incorporate rheology modifying additives to produce the thixotropic viscosity necessary for application still tends to give less satisfactory flow and levelling than is available in solvent-borne paints.

Latex-based paints for architectural use have been available for 50 years and today it is probably true that some 90% of both interior and exterior trade sales paints are water-borne. There is a significant market for water-based coatings in packaging applications such as overvarnish where currently solvent or high solvent content systems are used. Latexes may well find a good outlet in this area.

Although there are applications for latexes in heavy duty industrial coatings and maintenance paints, these are as yet not large volume. However, with increasing legislative environmental control the growth in latex use is likely to increase enormously.

Considerable effort has been expended in improving the polymer properties by the application of sequential emulsion polymerisation and the improvement of poly(vinyl acetate) by co-polymerisation with monomers such as vinyl esters of longish chain (C_8–C_{12}) acids or highly branched acids of similar carbon content, *e.g.* 'Versatic' acids (Shell Chemical).

Table 8.4 gives some idea of the types of latex used in different applications.

Table 8.4 *Some latex resins and their applications*

Application	*Polymer latex*
Interior flat paints	Styrene–butadiene, vinyl acetate, acrylic copolymers
Interior semi or gloss	Vinyl acetate, acrylic copolymers
Exterior paints	Vinyl acetate, acrylic copolymers
Industrial paints	Styrene–butadiene, acrylic copolymers
Speciality crosslinking paints	Acrylic esters

Pseudo-latexes

The utility of polymer latexes prepared from monomers by emulsion polymerisation has already been mentioned but it is obvious that not all polymers may be prepared in this fashion. It would be of considerable value to have polymers that fall into this category in a latex form, and particularly with the particle size in the range of a typical latex, *i.e.* 0.1–0.5 μm.

Such polymers may be dissolved in suitable solvents and emulsified in water, the solvent later being distilled out. This procedure is limited by the fact that the particle size of the polymer is very often $>1\ \mu$m which leads to instability regarding sedimentation. An interesting advance in this area was made by Vanderhoff *et al.* when they disclosed a method by which latexes of a variety of polymers, unavailable by emulsion polymerisation, could be prepared where the particle size of the polymer was very similar to that achieved in conventional procedures.

The method is based on the observation that in the preparation of monomer emulsions very small droplet sizes may be obtained when a mixture of a conventional oil-in-water emulsifier and the alkane or alkanol corresponding to the hydrophobic part of the emulsifier is used, *e.g.* hexadecyl sulfate and hexadecanol. In the pseudo-latex procedure, the polymer is dissolved in a suitable water immiscible solvent and the solution emulsified in a water solution of the emulsifier and co-emulsifier. Homogenisation with a commercial homogeniser is carried out until the droplet size of the disperse phase is 0.1–0.5 μm. Solvent may now be stripped out to leave a polymer latex with water as the continuous phase.

A typical recipe, taken from the patent covering this procedure, is given here.

Following the emulsification, homogenisation, and solvent stripping, there is obtained a pseudo-latex containing 45% epoxy resin with a particle size $\leq 0.3\ \mu$m.

Epoxy resin, *e.g.* Epon® 1001	31.25 g
Toluene–methyl isobutyl ketone (1 : 1)	93.75 g
Hexadecyltrimethylammonium bromide	0.78 g
Cetyl alcohol	1.73 g
Water	375.00 g

DISPERSION POLYMERISATION

In the section dealing with emulsion polymerisation we were considering a compartmentalised system of monomer dispersed in an immiscible liquid containing a soluble initiator. Let us now turn to the situation where the polymer is obtained as a colloidal dispersion in an organic liquid. Such systems have been developed, chiefly by ICI, and although the original non-aqueous dispersions (NADs) have tended to become unattractive due to environmental legislation against volatile organic solvents, the technique may be used as a step in the preparation of water-based coating. The Aquabase® technology developed by ICI for water-borne coatings in the automotive industry is an example and over 1 000 000 cars have been coated with such products.

It is interesting to note that dispersions of crosslinked hydrophilic polymers have been prepared directly from water-soluble monomers and a crosslinking agent, using ^{60}Co radiation and a polymeric stabiliser.

Aqueous polymer colloids formed by condensation polymerisation have been reported and, for example suitable modification of glyoxal–melamine polymer gives a product useful as a coagulant.

Technical Aspects

Only the non-aqueous system will be considered and we shall briefly present the mechanism of stabilisation whereby lyophobic colloids are stabilised in low polarity media.

A solution of the chosen monomer is prepared in a suitable solvent, usually an aliphatic hydrocarbon, that contains a block or graft co-polymer which will act as the colloid stabiliser. Under anaerobic conditions, with stirring, a suitable initiator is added and the polymerisation reaction allowed to continue to completion. The product will be a dispersion of polymer which may, under certain conditions, reach a concentration of 80%. Particle size may be varied as required and can be as small as 0.01 μm. The viscosity of a typical NAD with 60% dispersed polymer is around 0.05 Pa s at room temperature.

The polymeric stabiliser is an essential feature of this type of polymerisation and may be a block or graft polymer in which one component has a strong affinity for the polymer colloid whilst the other component has a strong affinity for the continuous phase. Such polymers having differing affinities to the two phases are said to be amphipathic and the stabilisation is termed steric although, the less used, entropic is more accurate. Structures of block and graft copolymers are given in Chapter 7. A particularly recommended modification of the graft structure is the comb type that gives excellent results (1).

```
M  M  M  M  M  M  M  M  M
|  |  |           |  |  |
M' M' M'          M' M' M'
|  |  |           |  |  |
M' M' M'          M' M' M'
|  |  |           |  |  |
M' M' M'          M' M' M'
```

(1)

In these structures, we may look upon the M component as the part that associates with the polymer particle, whilst the M′ part is well solvated by the continuous phase and is extended into it. The M part is usually known as the anchor component and may be connected to the polymer by physical attraction, covalent bonding, or incorporation into the particle.

Looking at the situation in a qualitative way, we may say that stabilisation results from the repulsive forces which occur from loss of configurational entropy when the particles approach one another and the solvent chains of the stabiliser start to compress or overlap. This loss in entropy is sufficient to overcome van der Waals–London forces of attraction so that the colloidal polymer is stabilised against flocculation, the steric barrier formed by the well extended and solvated part of the stabiliser preventing close contact.

An interesting application of NAD technology has been in the preparation of micro-gels which will form a later section of this chapter. Although, very largely, the technique has been used for radical polymerisation, it is also applicable to anionic mechanisms. Block polymers of t-butyl styrene and styrene have, for example, been shown to act as efficient stabilisers for the anionic polymerisation of styrene.

This very brief outline of what is a thoroughly studied topic cannot hope to convey the breadth of this subject from both theoretical and practical considerations. The definitive text is 'Dispersion Polymerisation in Organic Media', see the bibliography for details.

WATER REDUCIBLE SYSTEMS

Although, very frequently, referred to as water-soluble polymer systems, the water reducible system is not in fact a true solution, but rather a well solvated collection of aggregates. The principle of the method used to attain water reducibility is quite simple. The polymer, epoxy or alkyd for example, is chemically modified so as to impart acid or amine functionality, the reaction usually being carried out in an organic solvent. The function introduced is neutralised with base or acid and the salt so produced allows dilution with water. The polymer aggregates, swollen with the water-solvent continuous phase, are usually at a concentration of around 20% and the particle size is typically ≤ 0.01 μm, hence these systems are transparent or translucent. A limitation of these 'pseudo solutions' is that their viscosity is dependent on the molecular weight of the 'solute'.

In order to conform with legislation regarding volatile organic compounds, some effort has been directed to the preparation of water reducible resin systems without the use of amine or co-solvent. Blank has advocated the use of non-ionic groups, *e.g.* polyol or polyether, that may be introduced into polymer structures. There are certain disadvantages, such as poor water resistance and adhesion to substrates, but it is possible to optimise performance to acceptable levels.

Polymer modification with zwitterionic functionality has been recommended by Wicks and his co-workers. In this procedure both the carboxylic acid and amine are covalently bonded to the polymer.

There is a variety of chemical routes available and we shall merely mention four:

(i) maleated epoxy resin with dimethylaminoethanol;
(ii) styrene maleic anhydride with dimethylaminoethanol;
(iii) aryl cyclic sulfonium zwitterionic monomers; and
(iv) epoxy resins esterified with phosphoric acid with dimethylaminoethanol.

An interesting use of water reducible resins is in their use as stabilisers in emulsion polymerisation procedures. They may also be used at higher concentrations and become part of the polymer film. Use as modified polyester in emulsion polymerisation is exemplified in the section on micro-gels.

An important use of water reducible resins, particularly epoxides, is in the packaging coatings industry where wash coats are predominantly of this type. Reaction of an epoxy resin with phosphoric acid, advancement of molecular weight, and neutralisation of acid function, produced by hydrolysis of the hydrolytically unstable di- and tri-esters, is a common route to the wash coats.

Electropaints

Another important use of water reducible polymers is in the electrocoat field, particularly as primers in the automotive industry. The principle of the method lies in the susceptibility of lyophilic colloids to coagulation in the presence of electrolytically generated ionic species, leading to deposition of the polymer on to the workpiece, which may be made the anode (anaphoretic systems) or the cathode (cataphoretic systems). Typical reactions are given in Scheme 1.

ANODE

$$2H_2O - 4e^- \longrightarrow 4H^+ + O_2$$

$$\text{Polymer—COO}^- + H^+ \longrightarrow \text{Polymer—COOH}$$

CATHODE

$$2H_2O + 2e^- \longrightarrow 2OH^- + H_2$$

$$\text{Polymer—}\overset{+}{N}R_2H + OH^- \longrightarrow \text{Polymer—NR}_2 + H_2O$$

Scheme 1

Equivalent reactions at the opposite electrode lead to regeneration of the neutralising species which may be low molecular weight amine or acid respectively. These can be segregated from the bulk by semipermeable membranes or allowed to mix with the bulk and removed at intervals by dialysis or ultrafiltration.

Anodic electropaints are very often polymeric carboxylic acids neutralised with tertiary amines, although other anionic groups and inorganic bases have been used. Backbone structures that have found favour include acrylics and maleinised fatty acids, epoxy esters of fatty acids, and polybutadienes. Although performance of anaphoretic paints can be very good, they suffer from a lesser degree of corrosion resistance as compared to cataphoretic paints. This is due to the participation of a steel substrate in an electrochemical reaction outlined in Scheme 2. This

$$Fe - 2e^- \longrightarrow Fe^{2+}$$

$$2\ \text{Polymer—COO}^- + Fe^{2+} \longrightarrow \text{Polymer—NR}_2 + H_2O$$

Scheme 2

results in a disruption of the substrate surface and dissolution of iron may promote corrosion.

Cataphoretic paints are not prone to this problem and they are very widely used. Aminated epoxy resins are popular in this area and are neutralised with acetic or phosphoric acid. Blocked isocyanates are the most frequently used crosslinkers. They form a stable dispersion along with pigments and other additives in the presence of the dispersible film former.

Depending on the voltage used and a balance of the resistance of the liquid paint with that of the coated parts of the workpiece, an electrocoat is 'self-limiting' in thickness and can 'throw' into enclosed spaces such as box sections and car doors. As a result, unparalleled corrosion protection to objects of complex shape is attainable.

PLASTISOLS

A particular type of dispersion that has wide application in the coatings industry is that of poly(vinyl chloride) and/or poly(vinyl chloride) co-polymers, with, *e.g.* diethyl maleate, for flexibility, in compatible plasticisers. Such dispersions are known as plastisols, a term which should be reserved for resin–plasticiser systems. These, however, are usually too viscous for coating and adjustment is made by addition of volatile solvent, the resulting product being known as an organosol.

PVC plastisols have been prepared and used for nearly 50 years, the term plastisol having been first used in 1946. The partial crystallinity of PVC seems to allow the formation of stable dispersions in plasticiser and subsequent solvent dilution without causing excessive swelling. Attempts to use this technique with acrylic resins has not been successful so far, but there would be a very useful system for high-solids coatings formulations if small sized acrylic particles could be modified or placed in an environment where swelling could be prevented.

A relatively recent development, that is somewhat akin to plastisol formation, lies in the fluoropolymer area where Asahi Glass Co. have investigated poly(vinylidene fluoride). The excellent weather resistance of fluorinated polymers is well known but the high temperatures required for acceptable application, *i.e.* thin and pinhole free, limits the use of many thermoplastic fluoropolymers. By using acrylic resin solutions with PVDF, melt-flow properties are improved and the blended system has found use in coil-coating. The poor adhesion of PVDF to substrate is overcome in the acrylic blend.

Poly(vinyl fluoride) has a crystalline structure that with solvents at elevated temperatures forms smooth flowing systems that give defect-free coatings. In this case, as well as for PVDF, the solvents used should be looked upon as latent since they are only effectual when the resin–solvent system is hot.

The advantage of the plastisol approach is that it overcomes the limitations of solutions, where viscosity constraints determine the solids contents. With present restrictions on organic solvents, the use of plastisols is likely to diminish, unless some means is found whereby resins of varied constitution can be formulated into very high (>85%) solids systems.

MICRO-GELS

A form of polymer that is readily formed as a dispersion by non-aqueous or emulsion polymerisation is that known as micro-gel. Such polymers have been found to have useful application in the coatings industry and their increasing use warrants a brief discussion.

Dispersions of small particles of intra-molecularly crosslinked polymer particles having sizes $<\sim75$ nm may be considered as true molecules and as such may be treated by the thermodynamic equations that are applicable to polymer solutions. Material of this type was first prepared as long ago as 1934 and its adventitious presence in alkyds recognised around 1949 when the term micro-gel was first used. Structural studies were subsequently made. Over the past 15 years, application to coatings, where their presence imparts enhanced film strength and rheological properties, have been strongly championed by, notably, W. Funke. In the world of industry, Nippon Paint and ICI have numerous publications relating to the synthesis and application of micro-gels, whilst Pittsburgh Plate Glass have many contributions to micro-gel technology related to the automotive sector. It may be remarked that recent developments in applications of micro-gels to catalysis and as phase transfer agents have tended to increase in commercial production, notably in Japan.

Micro-gels may be prepared by the crosslinking of linear polymers in dilute solution so that inter-molecular crosslinking is minimised. However, this is not an economic route, and polymerisation of crosslinkable monomer mixtures in a controlled volume, such as may be achieved in emulsion polymerisation or non-aqueous dispersion systems, is the favoured route.

Careful selection of emulsifier is needed for successful micro-gel synthesis and it is necessary to design the procedure so that the conditions will allow all the monomer to be contained in micelles. Conventional surfactants, *e.g.* sodium dodecyl benzene sulfonate, may be used but far better results are obtained when polymeric surfactants based on polyesters or epoxy resins are used. Zwitterionic structures appear to give the best results and considerable advances have been achieved by Nippon Paint in this area. Using an epoxy-based surfactant containing the zwitterionic group (2) micro-gels of acrylic or vinylic monomers having particle size 20–50 mm are readily obtained. Another useful surfactant is based on polyesters which are reacted with *N*,*N*-bis-2-(2-hydroxyethyl)-2-aminosulfonic acid (3).

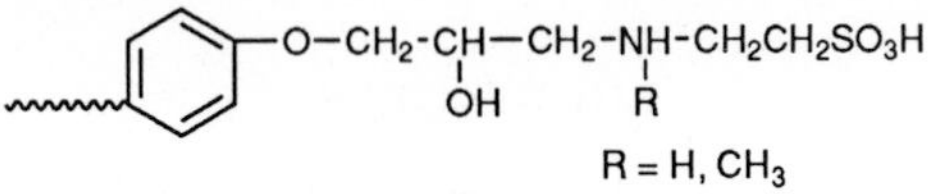

$$\sim\!\!\sim C_6H_4\text{-}O\text{-}CH_2\text{-}CH(OH)\text{-}CH_2\text{-}N(R)H\text{-}CH_2CH_2SO_3H$$

$R = H, CH_3$

(2)

$$(HOCH_2CH_2)_2N\text{-}CH_2CH_2SO_3H$$

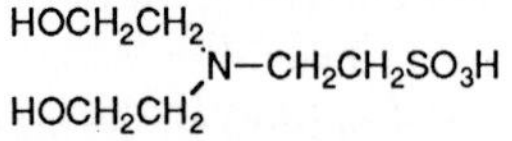

(3)

Specific applications in coatings lie mainly in the automotive industry where the micro-gel has been found to control metal flake alignment in metallic coatings. ICI's Aquabase system is based on micro-gel synthesis and has had considerable success in this area.

In water-based coatings, the presence of micro-gel assists in the water release whilst the low viscosities of micro-gel solutions, as compared to solutions of linear or branched polymers, makes their use attractive in high solids coatings. With solids contents of $>80\%$ in such coatings, the need for rheology modifiers that will lead to the required pseudoplasticity can be fulfilled by using micro-gels.

The use of suitably functionalised micro-gels as latent crosslinking agents may be an application having important consequences. For example, a carboxyl functional micro-gel in a carboxylated acrylic latex could be crosslinked by ions such as Zn^{2+}, Zr^{2+}, or Ca^{2+}.

Observations of increased mechanical strength in films of alkyd resins that contain micro-gel have been reported for other resins.

SUSPENSION POLYMERISATION

Although strictly not used for binder preparation for liquid paints, the process of suspension polymerisation may have application for the preparation of resins for powder coatings.

The process itself is very simple and may be considered as a water-cooled bulk polymerisation where the monomer is dispersed, in the form of small droplets in water. Although most commonly applied to free-radically initiated polymerisation, the method may also be applied to step growth mechanism types.

In a typical suspension polymerisation, a water insoluble or very slightly water insoluble monomer (or monomers) is mixed with a monomer soluble free radical source and the mixture dispersed in water that contains a suspension stabiliser. The degree of agitation largely determines the droplet size and when the desired size is attained polymerisation may be initiated. Other variables that influence the particle size of the final polymer are the dispersant, the viscosity of the system, and the interfacial tension between the monomer and the continuous phase.

A very large number of suspension stabilisers have been suggested for suspension polymerisation and these may be organic polymers such as poly(vinyl alcohol), poly(vinyl pyrrolidone), poly(acrylic acid), carboxymethyl cellulose, or inorganic compounds such as talc, bentonite, or freshly precipitated calcium phosphate.

An important parameter in suspension is the particle identification point which is defined as that percent of monomer conversion after which the polymer particles retain their identities for the remainder of the polymerisation. An idea of this may be attained even in an unsuccessful first run after which changes in initiator or temperature may be made to adjust the polymerisation rate.

An example of a suspension polymerisation taken from the literature is given as being fairly typical regarding monomer–water ratio, and initiator and stabiliser concentrations.

Water (200 ml) and poly(vinyl alcohol) (0.25 g) are warmed in a multi-necked flask fitted with nitrogen sparge, stirrer, and thermometer until the PVA has dissolved. At 20 °C a solution of 2,2′-azobis(2-methylpropionitrile) (0.5 g) in *p*-vinyl aniline (50 g) is added to the vigorously stirred PVA solution and the agitation continued for 30 min. Gentle stirring is now used whilst heating to 50 °C, at which temperature the polymerisation is allowed to proceed overnight. The polymer beads so formed are filtered and washed with water.

BIBLIOGRAPHY

Emulsion Polymerisation

D. C. Blackley, 'Emulsion Polymerisation. Theory and Practice', Applied Science Publishers, Barking, 1975. Now outdated but remains an excellent and clear exposition of the micellar mechanism of emulsion polymerisation.

'Emulsion Polymerisation', ed. I. Piirma, Academic Press, New York, 1982. A multi-authored collection which updates Blackley's text.

Dispersion polymerisation

'Dispersion Polymerisation in Organic Media', ed. K. E. J. Barrett, John Wiley, Chichester, 1975. The definitive text on the subject by those who pioneered it.

Suspension polymerisation

'Polymer Processes', ed. C. E. Schildknecht, John Wiley, Chichester, 1977, pp. 106–142.

F. H. Winslow and W. Matreyek, *Ind. Eng. Chem.*, 1951, **43**(5), 1108.

Various aspects of polymer dispersions

'Future Directions in Polymer Colloids', eds M. S. El-Asser and R. M. Fitch, NATO ASI Series, Series E, Applied Sciences No. 138, Martinus Nijhoff, Dordrecht, 1987.

Chapter 9

Formulation of Coatings Compositions

K. F. BAXTER

INTRODUCTION

Chapters 7 and 8 have described the central function of polymers in the chemistry of film formation. However, a successful coating usually requires properties which cannot be provided by any one component alone. The role of the paint formulator is to bring together the required constituents in a stable, cost-effective composition which can be conveniently applied to the substrate. No small challenge!

A paint film consists of a dispersion of a pigment or a mixture of pigments, extenders, *etc.*, in a binder or polymer. Other materials may be present to achieve specific properties. They may be organic solvents or water to give the required viscosity, suspending agents to keep the paint in good condition during storage, driers and accelerators which provide for rapid cure of the polymer, flow agents, anti-cissing agents, and so on. For the considerations which are to be expanded in this Chapter we will assume for simplicity that a paint film consists of pigments/extenders, polymers, and sometimes solvent.

PIGMENTS AND PIGMENTATION

Pigments have been used by man since the birth of civilisation. Coloured minerals and materials such as charcoal, were used to colour the bodies and living spaces of primitive man. As his skill grew, the use of minerals was to include the colouring of pottery, ceramics, and eventually glass, and the preparation of mixtures with media such as natural oils, and other resinous materials, to make paints.

Without the highly durable pigments which our ancestors used in their relatively simple paints, we would not now be able to understand so fully the development and growth of civilisation. From these early beginnings of the use of paint for decorative and artistic purposes has come the highly technical and commercially important paint industry of today.

So what is a pigment and what is a dye, and why are they different?

A pigment is defined as a coloured or non-coloured, black or white, particulate compound which can be dispersed in a medium, resin, or polymer, without being dissolved or appreciably affected chemically or physically.

When a paint is applied as a thin film over a substrate, the dispersed pigment will absorb and scatter light. This is one property which makes them so important in paint formulations.

Dyes or dyestuffs, on the other hand, are usually soluble in paint media and give transparent or translucent films. This property obviously limits dyes' utility in the coatings industry to products such as inks and stains.

Properties of Pigments

Other properties of pigments which are of great importance in coatings are as follows.

Colour. The primary use of coloured or black and white pigments is to produce films which when applied to substrates such as metal, wood, or concrete give the substrate a uniform distinct colour. The colour which is selected for aesthetic reasons must be durable, without fading or darkening, and the film must last for some considerable time when exposed to a wide range of different environments.

Opacity. To give a uniform colour in a minimum thickness, the pigment must scatter, reflect, or absorb light to prevent it reaching the substrate. White pigments opacify media by scattering and reflecting light, whilst coloured pigments also rely on the absorption of light to give colour and opacity. The physical property of pigments which controls the degree of opacity is the difference between the refractive index of the pigment and that of the medium.

Mechanical Properties. Depending upon the pigment's physical properties (such as hardness) and the amount dispersed in medium, tough, hard, yet flexible films which resist abrasion, impact, and other mechanical insults can be formulated. This is particularly important for coatings, for example on the internal and external areas of ships, oil rigs and platforms, storage tanks, the under-parts of vehicles, and the exteriors of aircraft.

Durability. Pigments play an important role in the protection of the paint media when films are exposed to atmospheric weathering. They can absorb or reflect UV radiation which would otherwise cause the breakdown of polymer systems. This can be shown in the comparison of the durability of clear films with the durability of pigmented films of polymers such as alkyds. The clear films lose gloss and degrade much more rapidly.

Barrier Properties. Pigments also improve the water resistance of paint films. Depending upon the particle shape, and the amount of pigment in the coating, the passage of water molecules and other ions through the film can be restricted to very low levels. Lamellar (plate-like) pigments such as aluminium flake, mica, micaceous iron oxide, and others are particularly useful in coatings to protect substrates immersed in water for long periods or subject to extreme weather conditions. Such are ships, bridges, and other structures.

Anti-corrosive Properties. Almost all substrates used in industry will deteriorate if left exposed to natural environments for any length of time. This applies to ferrous and non-ferrous metals, concrete, and wood. Only in exceptional circumstances and locations can such substrates be left unprotected. It is desirable, if not essential, that all industrial substrates be adequately protected so that the structures maintain their strength and performance..

Iron, steel, zinc, and aluminium can be protected by inhibitive pigments in media. Without such protection the many technical and industrial uses of these metals would not have been possible. Ships, vehicles, aeroplanes, bridges, and industrial plant would not work for long if they were not given the protection afforded by anti-corrosive paints.

Metallic pigments can act sacrificially as the anode in the corrosion mechanism and thus prevent electro-chemical dissolution.

Reinforced concrete will corrode and the structures degrade if they are not protected by special pigmented coatings. Spalling of the concrete overlayer caused by the expansion of corroding steel reinforcement is prevented by coatings which delay the ingress of water, carbon dioxide, and other ions into the concrete substrate. This can be achieved by understanding the relationship of the selected pigments with the media used for these coatings. Likewise the coating of the steel reinforcement with anti-corrosive systems preserves concrete structures.

Biocidal Properties. Special pigments, for use in antifouling and fungicidal coatings, have attained very valuable industrial application. They are available in the paint film to give an environment which is hostile to the growth of disfiguring algae, animal species, and fungi on the immersed hulls of ships, the exterior of buildings in tropical areas, and industrial plant such as food factories and breweries, where conditions are conducive to the growth of algae and fungus.

Chemical Resistance. The chemical resistance of paint films can be improved by the judicious use of pigments. This also applies to coatings which have to resist the degrading properties of solvents, oils, fats, acids, alkalis, and other chemicals.

Fire Resistance. Fire resistance of coatings is provided by pigments which cause the film to intumesce. This expansion and the char which is produced when the protective layers are exposed to very high temperatures insulate the structure for a time to allow the evacuation of buildings, and also to give time for the destructive fire to be fought, and the construction to be saved.

Rheological Modification. The structures of liquid paints play an important role in their storage and application properties. It can prevent the settlement of paint ingredients over long periods, and also allow the application of thick films of paint without the unsightly sagging, running, and excessive flow which can happen with coatings which are not formulated correctly. Any dispersed particulate material can effect the viscosity of a composition, but certain pigments have a profound effect on rheology (*c.f.* Chapter 4).

Types of Pigment

Historically, the pigments first used in paint were essentially minerals produced from natural deposits. These were mined, crushed, and ground, sometimes washed, and then chemically modified as knowledge and inventions improved their properties.

Coloured pigments took names of the area where they were first discovered and processed, and indeed to the present time names such as Sienna, Venetian Red, Umber, and Spanish Ochre are used to describe a colour or property of a particular pigment.

Today many compounds are still mined from natural sources and treated to give pigments for paint and other uses. The vast majority of such pigments, however, are nowadays the so-called extender pigments. These extenders, or fillers, as they are sometimes described, play a very important part in the formulation of paint, and although they are less costly than coloured pigments or other special pigments, their total contribution to the properties of paint must not be ignored.

The 'prime' pigments, coloured pigments, anti-corrosive pigments, and other manufactured pigments, are produced by the chemical industry as inorganic metallic salts, metallic, and non-metallic organic compounds. The organic compounds, in general, have come from the dye industry which has developed and is developing dyes for the textile and wool industries. The manufacture of pigments is now a massive world-wide business which is of huge economic importance.

In order to understand the role of pigments in coatings formulations, it is necessary to classify the pigments into groups and sub-groups but before we can do this we must consider the use of each particular layer of paint or coating in a protective coating system.

In simple terms, a typical coating system might be comprised of a primer, a build coat, and a finishing coat. There are variations of the theme where primer and build coat are formulated as one, or where the build coat and finishing coat are formulated as one coat, but for simplicity we will accept the three coat system as ideal. How do we formulate these coatings?

The first coat, or primer, must offer protection to the substrate, give a uniform opacifying colour, and above all give good adhesion to the substrate and offer adhesion to the subsequent coats of paint. If they are needed, anti-corrosive pigments are included in the primer. As these control the corrosive processes at the substrate they must be as near as possible to the surface of the metal they are protecting. It would be a pointless exercise to formulate build coats or finishes with the anti-corrosive pigment as this would be a waste of expensive pigment.

Primers may also be formulated using metallic pigments which act as sacrificial anodes.

The primer should also act as a barrier coat if possible. The barrier effect slows down the ingress of water molecules and other ions. This is attained by the use of extender pigments which are chosen because of their hydrophobic and lamellar properties. The colour of primers is not important, except that it must be uniform and distinct from the substrate. Less costly pigments, such as iron oxide, are widely used as colouring pigments for primers.

The build coat in a protective coating system gives thickness to the system. It helps to eliminate imperfections in the surface being coated so that the minimum film thickness which is necessary to give long term protection is obtained. The optimum thickness of a build coat largely depends upon the environmental conditions and the structure being coated, but the thickness should be at least that which gives a uniform film, free of imperfections, free of pinholes, and which can fully coalesce.

The build coat also provides additional barrier properties giving maximum impermeability. This is decided by the type of extender pigment and the volume of pigmentation in the film. The build coat also contributes to a uniform colour and good opacity.

The finishing coat of a system gives the decorative effect, and as it is the coat which is constantly 'in the eye' the effect given must be satisfactory in every way. The colour and gloss must be correct, must be uniform, and should not change over a long period of time.

The coloured pigments used in the finishing coats are those which have proven opacity and durability. They are used at the lowest effective concentrations, bearing in mind that most coloured pigments are very expensive. Extender pigments are also used in finishing coats to control flow, improve the toughness of the film, and lower the gloss of the finish if this is required.

The whole system applied as three coats must retain its integrity without loss of adhesion, without blistering, with good colour retention, and above all it must protect the substrate from degradation.

In the Protective Paint Industry pigments are classified into the following groups:

Anti-Corrosive Pigments. As the name implies, these materials are used to prevent the corrosion of the substrate. For ferrous metals there are two ways in which this can be done, either by inhibiting the corrosion reactions or for the pigment to act as a sacrificial anode, which protects the substrate which itself becomes the cathode in the corrosion cell.

Pigments which act as sacrificial anodes are essentially small particles of zinc metal or alloys of zinc of 4 to 7 μm in size. They are formulated as a range of zinc rich primers with media such as inorganic and organic silicates, epoxy, chlor rubber, and other inert polymers. The essential point about these formulations is that the zinc content of the film must be high so that life of the primer is long, that is, there is sufficient zinc available to be 'sacrificed' to protect the steel substrate.

As this is an electrochemical process there must be metallic contact so that electrons can pass easily. It is achieved by high zinc concentrations.

It has been suggested by many workers that the zinc content of primers should be in excess of 90% by weight in the dry film to give good long term protection. The actual concentration of zinc metal used in zinc primers depends upon the particle size of the zinc and the media used in the primer formulation, and the skill and knowledge of the formulator in giving viable, long term economic products.

For ferrous and non-ferrous metals, the corrosion processes can be controlled by inhibitive compounds, which slow down or prevent the corrosion cell mechanisms from degrading the metal substrate. The way in which inhibitors act is very complex and cannot be dealt with in this review. All that can be said is that the anti-corrosive pigments passivate the substrate and the corrosion rate is reduced to very low levels.

Over the years a considerable number of inhibitive pigments has been developed, marketed, and used by the paint industry. For ferrous metals these include oxides, chromates, phosphates, molybdates, borates, and complexes formed from these families of compounds. Most are salts of lead and zinc with some strontium, calcium, and barium compounds.

For non-ferrous metals the most useful inhibitive pigments are chromates, mainly of zinc, but barium and strontium chromate have been used.

Before abrasive blasting techniques became established as the normal treatment of steel used for the building of structures, tanks, ships, bridges, *etc.*, the method of removing millscale, the oxide layer on new steel, was to allow the steel to corrode. The millscale became detached, leaving a rusty surface which was then wire brushed or flame cleaned, leaving steel with an adhering layer of oxide on the surface. It was to this surface that priming paints were applied. For these conditions the most widely used inhibitive pigment was lead sub-oxide, Pb_3O_4, known as red lead. It was found that this pigment when dispersed in a medium, essentially linseed oil, gave excellent protection to steel cleaned by the processes described above.

The virtue of a red lead in oil primer is that the lead oxide reacts with fatty acids present in linseed oil to produce soaps which help in giving water resistance. The paint, based on drying oils, has excellent wetting properties, which ensures that the rusted surface is saturated with the priming paint. It is also thought that ions such as sulfates, which take part in the corrosion reaction cycle, are removed by lead oxide to form lead sulfate, and thus the corrosion process is stifled. Adding all of the possibilities together, well formulated red lead in oil primers are very effective in preventing the corrosion of steel which is subjected to normal weathering.

However, as with most things, it has disadvantages. Red lead has an extremely high relative density, which leads to severe settlement problems on storage of paint. It is also difficult to apply on a large scale. Linseed oil is very slow in drying, which leads to overcoating problems and, due to the reaction to form soaps, thickening and gellation can occur on storage. The most important disadvantage is that red lead, like all lead compounds, is considered a toxic material and its commercial use is discouraged, sometimes by legislation, in many countries.

Red lead was a very important pigment, it was widely used, and this prompted the development of many other pigments to overcome its deficiencies and disadvantages.

Lead silicochromate was widely used, especially in the USA. It acts in a similar manner to red lead and has a lower relative density. It consists of lead chromate particles coated with silica to control their rate of dissolution. It has, nevertheless, fallen into disfavour since it contains lead and hexavalent chromium.

Calcium plumbate is a pigment developed particularly for use in primers for galvanised steel but has suffered the fate of all lead based anti-corrosive pigments.

Zinc molybdate, calcium zinc molybdate, and zinc molybdenum phosphate are a group of compounds which have been promoted and used as anti-corrosive pigments. These, together with compounds such as calcium borosilicate and barium metaborate, are free of the toxic hazards associated with lead, and thus have a certain appeal as red lead replacements.

Chromates of zinc have also been used as anti-corrosive pigments for steel but they are more effective as inhibitors for aluminium and other non-ferrous alloys.

The most useful and manageable pigment developed as an anti-corrosive pigment for steel is zinc phosphate, $Zn_3(PO_4)_2.2H_2O$. It was shown that zinc phosphate in a drying oil medium offered long term protection to steel and because of the many advantages over red lead it has become widely used in the paint industry, especially in Europe.

Zinc phosphate is easily prepared from relatively cheap raw materials, waste zinc metal and phosphoric acid. It has low toxicity, it is white, and thus has the advantage of being able to produce light coloured primers, and it has a low relative density. The crystalline shape is lamellar and it acts as a barrier pigment. It can be formulated with many types of polymer as priming paints, and is quite stable in media such as alkyd, epoxy, chlor rubber, vinyl, urethane, and water-based systems.

The anti-corrosive properties of zinc phosphates are somewhat difficult to prove from classical studies but because of its barrier properties, colour, stability, and relative low cost, zinc phosphate has achieved predominance in the market.

Many attempts have been made to improve the anti-corrosive properties of zinc phosphate by mixing it with materials such as zinc molybdate, or co-precipitating it with this and other compounds.

Aluminium zinc phosphate, produced by co-precipitation techniques, is claimed to have improved inhibiting properties due to the higher phosphate content and higher solubility. The higher activity of zinc phosphate molybdate may be due to the molybdate ions.

Organic modification of zinc phosphate is also possible, and compounds have been introduced as improved anti-corrosive pigments. Nevertheless, few alternatives have gained a leading position in the market place.

The search for more effective anti-corrosive compounds goes on. The most interesting approach of recent years has been the development of ion exchange compounds, which remove ions such as chloride and sulfate which are necessary for the corrosion process to proceed.

For the protection of aluminium, and its alloys, basic zinc potassium chromate, $3ZnCrO_4Zn(OH)_2K_2CrO_4$; zinc tetroxychromate, $ZnCrO_4 \cdot 4Zn(OH)_2$; and strontium chromate, $SrCrO_4$ are used with various media. Strontium chromate in a cured epoxy binder has particular use in the aircraft industry.

Zinc tetroxychromate is used in the formulation of wash primers. These are based upon polyvinylbutyral (PVB) and phosphoric acid, the latter being added just prior to the application of the primer. Films of wash primers are applied, only a few microns in thickness, and these adhere extremely well to ferrous and non-ferrous metals. They form the base for other primers which would not necessarily adhere well to the metal surface. The roles of phosphoric acid, PVB, and chromate pigments in giving the very adherent films are not clearly understood, and many attempts to formulate competitive compositions as wash primers have failed.

Barrier and Extender Pigments. The binders or polymers used in protective coatings dry or cure at ambient temperatures either by solvent evaporation, air oxidation, or by chemical interaction of components which are mixed together before application. The morphology of these polymers is complex, and although they form continuous films they are invariably porous, to a greater or lesser degree, to molecules and ions of oxygen, water, chlorides, sulfates, carbon dioxide, *etc.* In other words, these films are not entirely suitable as clear films for long term protection of substrates against corrosion. They have to be reinforced by pigments. Pigments which are particularly suitable for reducing permeability are those which have plate like structures, the lamellar pigments. They can be metals, oxides, silicates, *etc.*, and are known for their barrier protection.

Aluminium flake is manufactured from high purity metal by milling aluminium powder in a solvent such as white spirit, with stearic acid as a lubricant. The stearic acid coats the aluminium flake and causes the aluminium to 'leaf', that is to float to the surface of a paint to form a layer where the overlapping flakes are orientated parallel to the surface to give a bright film with high lustre. If an alternative lubricant such as oleic acid is used in the manufacture, non-leafing aluminium flakes are produced. These, when distributed in a polymer, remain in the body of the film and do not migrate to the surface.

Leafing aluminium flakes are used in finishing paints, whilst non-leafing aluminium is used in primers and build coats. The particle size of the flakes ranges from about 1 μm to 100 μm in the longer dimensions and 0.05 μm to 2 μm in thickness.

Aluminium flakes reduce the permeability of polymers by causing ions or molecules to follow a long, tortuous path to the substrate, as compared with a direct path through an unfilled polymer.

In finishing paints leafing aluminium flakes can be used with other colouring pigments but the predominant pigment is the aluminium, which also has the advantage of having excellent resistance to ultra-violet radiation. So-called metallic finishes are widely used in the Automotive Industry where their characteristic ability to show a different colour when viewed from different angles, the so-called 'flip' or 'flop', has a strong bearing on body design. Less pronounced relief can be introduced into body panels with the same aesthetic effect.

In primers and build coats, non-leafing aluminium flake is mixed with other pigments such as micaceous iron oxide, silicates, silica, and filler pigments such as barium sulfate. The optimum amount of aluminium pigment used in the film is usually determined from permeability studies.

Thin flakes of chemical resistant glass can also be used in a similar manner to aluminium flakes. They are also manufactured by a milling technique, the flakes tending to be larger than aluminium flakes. Glass flakes are formulated into coatings which are applied by specialised spray techniques to give very thick films which have excellent resistance to abrasion. They are particularly useful for the repair of the insides of large storage tanks, the legs and undersides of oil rigs, and the external submerged surfaces of barges and ships which have suffered steel loss due to corrosion and abrasion.

The media used for these coatings are essentially solvent free materials, two pack epoxy and peroxide cured polyester being preferred.

Glass flakes being of low density are not easily handled during manufacture of these compositions, and care has to be taken to ensure that the flakes are well distributed in the media and thoroughly 'wetted out'.

One of the most important pigments used in coatings for the protection of large structures is micaceous iron oxide (MIO). This material is specular haematite (approximately Fe_2O_3 90%), a neutral mineral originally found in England and Austria. Other sources have been found but the Austrian mine still gives the mineral with the best lamellar structure. The term 'micaceous' refers to the mica like structure of the iron oxide which gives flakes up to 100 μm in size and approximately 5 μm thick. MIO has unique properties in having excellent heat resistance, good chemical and water resistance, it does not degrade under the influence of UV, and thus protects the media of paint films. It also has a low oil absorption which is important in allowing high loadings in paint. MIO can be formulated with a wide range of binders such as oleo-resinous materials, alkyds, chlor rubber, epoxies; vinyls and urethanes to produce primers, build coats, and finishes. The only disadvantage it has in finishes is its dark colour but it can be mixed with coloured pigments and aluminium to produce attractive films which sparkle in bright sunlight.

Many structures, some going back to Victorian times, have been maintained with micaceous iron oxide based coatings since they were built, with excellent results. German and British railways have specified the pigment for their use for many years.

In the formulation of MIO based coatings the extensive addition of other, cheaper, extending pigments must be avoided as these detract from the orientation of the iron oxide. High loadings of MIO are essential but there are optimum concentrations for each medium to obtain films with the best impermeability. High film thicknesses must also be applied to give good orientation of the MIO. There is a grade of a synthetic micaceous iron oxide being developed for use in high performance coatings.

Probably because of the absence of local sources of MIO in the USA, the largest use has been in Europe, the Middle East, and Far East. In the USA mica has been used extensively as a lamellar pigment.

Mica, an aluminium potassium hydrated silicate, is found in several forms, the most important for the paint industry being muscovite. Mica has a high aspect ratio, *i.e.* diameter to thickness ratio, but has a high oil absorption which limits its loadings in paint films to relatively low levels.

Additions of mica to zinc-based primers have been shown to improve the performance of these products.

Other lamellar pigments which have been marketed for the paint industry include stainless steel flake.

There are many other minerals which have been used in paints for many years. These were called extenders, and were considered as cheap materials which reduced the cost of the finished paint.

With a more enlightened scientific attitude the so-called extender pigments are now considered to be of great value in giving improved water resistance, improved durability, easier application, and good storage stability to paints, as well as improving commercial viability by lowering costs.

Extender pigments are generally white naturally occurring minerals with very low solubility in water and preferably inert to the action of acids and alkalis. This does not exclude coloured minerals such as iron oxides which have use in primer formulations as extending pigments.

The minerals are won by mining operations and are crushed, cleaned, and segregated as compounds with particles from sub-micron to a few microns in size. The minerals cover a wide range of chemical compounds such as sulfates, oxides, and silicates. Carbonates of calcium and magnesium are used extensively in paints for decorative purposes but because of their solubility, especially in acid solutions, are not formulated into coatings which require high durability and high water resistance.

Magnesium silicate, sometimes associated with other minerals such as aluminium silicates and quartz, is found in many parts of the world as talc, the most widely used extender pigment. It can have a fibrous or needle like structure, be plate like, or amorphous, depending upon the source. Talcs have important use for giving good reinforcing properties and improving the impermeability of films. The fibrous structures give good storage properties to paint by preventing settlement of pigments, but care has to be taken to exclude asbestiform fibres because of health considerations (inhalation of asbestos fibres of a certain size can lead to asbestosis and lung cancer). Talcs are generally hydrophobic in properties, they have good colour, and are easily dispersed in paint media.

Barytes, the mineral form of barium sulfate, is an inert compound with a low oil absorption and as such can be used in high loadings in build coats and primers. It has a relatively low cost and finds use in chemical resistance coatings. The precipitated form of barytes is Blanc Fixe which, being purer, is used as an extender for coloured pigments. Barium sulfate has the disadvantage of being high in relative density which can lead to settlement problems.

Synthetic silica, especially fumed silica prepared by flame hydrolysis of silicon tetrachloride, is an important product having an extremely high surface area and small particle size. Its high surface area gives a very high oil absorption and thus makes it suitable as a matting agent for coatings. It also gives structure to media to prevent settlement and to give non-sagging films when they are applied in high film thicknesses.

The extender pigments are formulated with anti-corrosive pigments and barrier pigments to give a wide range of products for the protective coating business.

Coloured Pigments. Colour, the measurement of colour, and an appreciation of the whole subject of the use and application of coloured pigments is a study in its own right, and can only be mentioned in a broad sense in a review such as this.

The colour of a paint film is controlled by several factors, apart from the colour of the pigment itself. The amount of pigment present has an obvious bearing on colour; a high gloss film will have a different colour from a low gloss film; the degree of dispersion of the pigment influences the colour of a film. The primary particle size of coloured pigments is small, generally less than one micron, but as these particles are present as agglomerates it is the extent of the breakdown of the agglomerates in the dispersion of the pigment which has an influence on the ultimate colour attained when a pigment is 'ground' into a paint binder.

The colour of a paint binder or polymer has an effect on the finished colour of paint. Acrylic media are more or less colourless, whilst some alkyds and oleo-resinous media are pale yellow when dry.

The properties of coloured pigments are briefly as follows.

Mass Colour—This is the colour given to a film by a pigment at high opacity.

Tinting Strength—This is the degree to which a pigment gives colour to a white base. A pigment with high tinting strength needs smaller amounts to give a colour compared with a pigment with low tinting strength. For a white pigment the converse applies.

Hiding Power—This is the property of giving a paint opacity, that is obliterating the colour of the substrate.

Light Fastness—The ability of the colour to remain unaffected when exposed to natural weathering. This is particularly important when coloured pigments are used to tint white paints to off-white colours. Pigments in the mass colour also may change colour on exposure to sunlight.

Durability—Coloured paints may appear to fade on exposure to UV radiation. This, however, may be entirely due to the oxidation of the polymer. Pigments can protect the polymer by absorbing the radiation and thus improve the durability of the coating. On the other hand, some pigments can speed up this degradation.

Bleeding—Coloured pigments can be soluble in solvents used in paints. This causes problems on storage as the pigment dissolves and then crystallises out of solution and the colour of the paint changes. Also, pigment of undercoats can bleed into the topcoat, changing the colour of the topcoat.

Chemical Resistance—Is also important, for example the resistance to acid rain, together with the reactivity of the pigment with media.

The number of coloured pigments used in the paint industry as a whole is quite large, but as the industry is split into several businesses it does not necessarily mean that the same pigments are used by all of the sub-sections of the industry. For example, for the automobile industry many pigments are specially developed and processed for that business. The following is a description of some coloured pigments used in protective coatings. For simplicity they are divided into colours—white, red, blue, yellow, and so on.

White Pigments. A white pigment is not a coloured pigment, in fact, it is a transparent material and appears to be white because of the small size of the particles of the pigment which causes light to be scattered and reflected back, and the eye receives the whole spectrum of reflected light. Best light scattering is obtained with the maximum difference in refractive index between pigment and medium. It is customary to think in terms of a highly refractive pigment, but it is also possible to produce a white coating by contrasting a medium with air introduced as pigment-sized glass bubbles. On the other hand, black pigments absorb light and a true black reflects no light. Coloured pigments absorb certain wavelengths and the eye only perceives the reflected wavelengths, which gives the colour of the pigment.

The white pigment which is used by the paint industry is titanium dioxide. It is the most important pigment used in the industry and has had an immeasurable affect on all aspects of decoration used by the industrialised nations.

Before the development of titanium dioxide in the 1920s and wide scale use in the 1950s, the white pigments used in paint were compounds such as lead carbonate, zinc oxide, antimony oxide, lead silicate, lead sulfate, and coprecipitated forms of these compounds.

Apart from zinc oxide in various forms, all other white pigments have been made obsolete. In any case, many would now be considered toxic and could not be used in most paints.

Titanium dioxide is manufactured in millions of tonnes per annum by two processes, the sulfate process and the chloride process, from natural ores found as ilmenite, an iron titanium dioxide, or rutile, a natural form of titanium dioxide.

In the sulfate process the ilmenite ore is reacted with sulfuric acid to produce titanyl sulfate which is hydrolysed to give hydrated titanium dioxide. This is then calcined to give the pigment titanium dioxide.

In the chloride process, rutile ore is treated with chlorine to produce titanium tetrachloride which is then vaporised and oxidised to produce titanium dioxide and chlorine which is reused in the process.

Two crystalline forms of titanium dioxide, anatase and rutile, can be made by the sulfate process. The chloride process only produces the rutile form. The rutile structure is a more compact tetragonal crystal which gives it better properties with regard to opacity due to its higher refractive index.

The rutile grade of titanium dioxide is coated with a surface layer of inorganic materials such as silica or alumina. Organic compounds, such as amines or silicone, can also be applied to improve the dispersibility and durability.

The rutile is generally preferred in high performance coatings.

Rutile titanium dioxide is almost an ideal pigment in that it is considered non-toxic, it can be used in foodstuffs, and it has the highest refractive index of all white pigments. It can be produced as a very small particle, approximately 0.2 μm, which gives excellent opacity. Its oil absorption is relatively low and its relative density is not too high permitting the manufacture of stable, fluid dispersions. The tinting strength of titanium dioxide is also good.

Its disadvantages are that it can cause degradation of some polymers by being photochemically active and that it does absorb some light at the blue end of the spectrum, which makes it a yellow toned white. For most purposes, however, this has little effect on the appearance of white paints.

Taking everything into account it appears that titanium dioxide is the 'ultimate' white pigment but it is interesting to note that white pigments do not exist in nature and that there are no organic white pigments.

Red Pigments. Red pigments can be divided into two groups, one a group which consists of various grades of iron oxide, whilst the other consists of organic compounds.

The readily observable differences between these two groups are that the organic compounds have a higher chroma and thus give brighter, purer, colours compared with the dull reds produced from iron oxide.

On the outskirts of the small village of Roussillon in Provence in Southern France, in an area of only a few acres, there can be observed and explored the residual piles and strata of a defunct mineral quarry. In this small area, in the bright Provencal sun, can be seen a glorious range of colours from the palest of yellows through bright yellow to orange, reds, pinks, browns, and almost black. Indeed, it is said that up to nineteen different coloured forms of iron oxide were mined from this small area until commercial pressures from other sources caused the mine to close. This is a fine example of the range of coloured pigments which can be obtained from one chemical compound by variations of particle size and minor variations of chemical composition. In reality, these bright-seeming colours are relatively dull, that is they have low chromas compared with other pigments such as the organic reds.

The iron oxides used as red colouring pigments are mainly 'synthetic' iron oxides which are produced by a number of processes. One process is direct precipitation. Calcination of synthetic yellow iron oxide, $Fe_2O_3.H_2O$, removes the combined water to give red Fe_2O_3. Calcination of black iron oxide, $Fe_2O_3.FeO$, which is produced synthetically also gives red forms of Fe_2O_3.

By careful control of the processes a range of red coloured pigments of varying particle size can be produced. These range from yellow toned to blue toned reds. They are also designed to give easy dispersion in media.

The properties of red iron oxide which make it important are the low cost, excellent light fastness, and heat resistance. It absorbs UV radiation to a certain extent which gives good durability, good hiding power, good chemical resistance, and it is insoluble in paint media and solvents.

The disadvantages are a poor tinting strength, for some purposes it has a dull shade but this is not so important when the decoration of large structures is considered.

For bright red coloured coatings a range of organic compounds is available, these are:

Toluidine reds [*e.g.* (1)] which are azo pigments manufactured by coupling, for example, diazotised *m*-nitro-*o*-toluidine with *β*-naphthol. They have very high chromas, producing bright colours which have good light fastness in the full tone but poor light fastness when reduced to tints. They also bleed in aromatic solvents. They find application in decorative paints which have general purpose use as, for example, air drying alkyd-based coatings containing white spirit solvent.

By changing the amine and introducing chlorine containing amines a range of products such as Para red (2) and Parachlor reds (3) can be produced. These have various shades of red and find some use in the paint industry but are very important in the ink industry.

Other red pigments which are used are Naphthol reds (4), where the *β*-naphthol is replaced by hydroxynaphthoic acid anilide in the coupling reaction with diazonium compounds. These again are used in decorative paints, especially in full shades.

The most recent red pigments introduced are the Quinacridone reds and violets [*e.g.* (5)]. They have excellent brightness, good durability, good solvent resistance, and have good tint stability. Unfortunately, they are very expensive.

Blue Pigments. The blue pigment which has universal application in the paint industry and which ranks with titanium dioxide as the most successful pigment in its class, is Phthalocyanine blue.

Phthalocyanine ('phthalo') blue, $C_{32}H_{16}N_8Cu$, (6) was discovered accidentally by observation in a chemical plant where phthalic anhy-

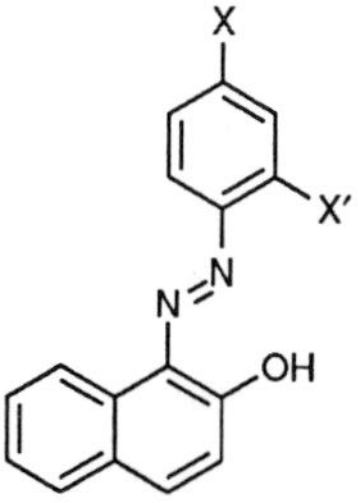

	X	X′
(1)	CH_3	NO_2
(2)	NO_2	H
(3)	Cl	NO_2

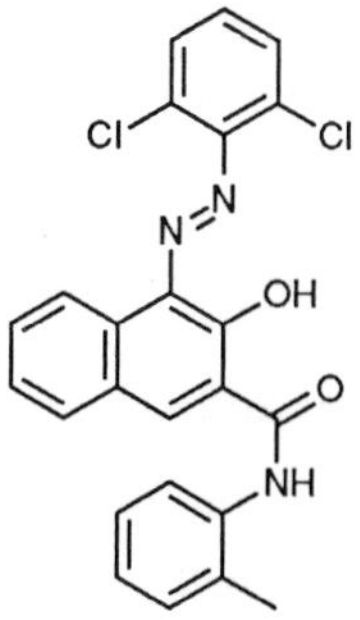

(4)

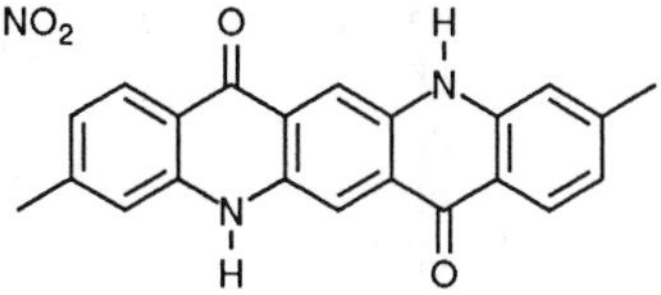

(5)

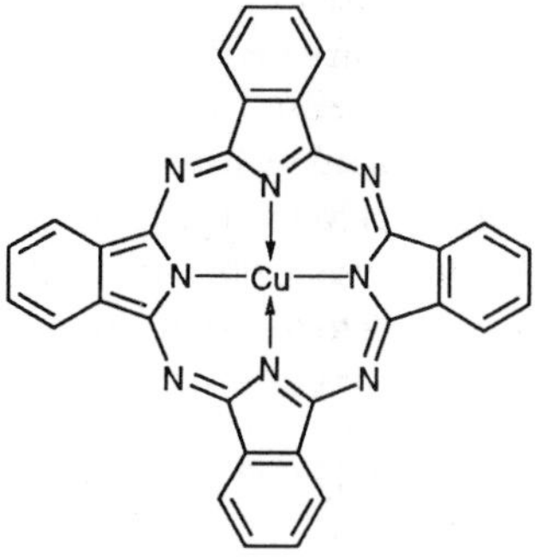

(6)

dride was being reacted with ammonia in a copper vessel. This discovery led to an industry which has proved to be very important for the paint, printing, plastic, rubber, and fibre manufacturing industries.

Phthalo blue is now made by reacting phthalic anhydride with urea, a copper compound, and a catalyst in a high boiling solvent. The crude pigment is then processed by one of two methods, acid pasting or salt grinding, to give a wide range of pigments of different particle size and crystal form. Copper free phthalocyanines and chlorine modification of the molecule are also possible to produce other grades of the pigment. The pigment can exist in a range of shades from red blues to green blues, in flocculating and non-flocculating forms.

The essential properties of phthalocyanine blue are its light fastness in full and reduced tone, its excellent chemical and solvent resistance, resistance to crystallisation, and flocculation in paints.

Phthalo blues are rarely used in the mass tone as they are rather dark but give superb colours when diluted with titanium dioxide and yellow pigments. They have replaced all other blue pigments such as Ultramarine blue (a sulfide of an aluminium silicate complex) and Prussian blue (potassium ferro-ferric cyanides) in virtually all coatings.

Green Pigments. The discovery of blue phthalocyanines gave impetus to the search for pigments with similar chemical structures but having other colours. This has resulted in chlorinated copper phthalocyanines and chlorobromo copper phthalocyanine pigments which are green in colour. These are commercially available and have similar properties to the blue phthalocyanines.

Green pigments are now dominated by phthalocyanine pigments either as green phthalocyanines or as mixtures of blue phthalocyanines with safe yellow pigments.

Cheaper greens formulated from blends of lead chromate and Prussian blue have more or less disappeared from the market due to the toxicity problem of lead compounds, but where possible blends of phthalo blue and yellow lead chromate are still used to produce various green colours.

One pigment which is green in its own right is chromium oxide, Cr_2O_3. This dull grey/green coloured pigment has excellent chemical resistance, good heat and light stability, good opacity, and is of low toxicity. It reflects infra-red radiation and finds application in deck paints and in camouflage coatings for military purposes.

Yellow Pigments. The yellow pigments are an important class of colours as they can also be mixed with other primary colours, reds and blues to produce orange, browns, and greens. As with the red pigments, they can be divided into inorganic compounds and organic compounds. The inorganic section contains yellow iron oxides, lead chromates, cadmium yellow (though this pigment is now considered toxic), and pigments such as zinc chromate, but this is mainly used for its anti-corrosive properties.

Yellow iron oxide is produced in a range of dull yellow shades synthetically from by-products of industry, scrap steel, *etc.* $Fe_2O_3.H_2O$ has an acicular or needle shaped crystal. The size of the acicular crystal determines the colour. Yellow iron oxide has all of the attributes mentioned for red iron oxide, except heat resistance since it loses the water molecule on heating to give red iron oxide. It is a relatively cheap pigment and is used extensively to produce cream tints with titanium dioxide, browns, and dull shades used in industry.

Lead chromates are manufactured by precipitating the pigment as lead chromate or co-precipitating lead chromate and lead sulfate to give primrose chromes, lemon chromes, and middle chromes which have a red/yellow tone.

Chromes are bright pigments with excellent opacity and are relatively cheap. They are fast to heat and are insoluble in solvents. They unfortunately contain lead and chromium and are not recommended for certain purposes in decorative coatings because of their toxic properties.

Lead chromates, if left untreated, darken on natural weathering. To improve this property lead chromes are surface treated with compounds such as silica and alumina. They are also affected by alkalis and can fade when exposed to chemical fumes. Nevertheless, as a group these pigments have been and are still important.

Associated with the lead chromes are mixed crystals of lead chromate and lead molybdate, the so-called scarlet chromes or molybdate orange pigments. They are available in colour range from orange to scarlet. The scarlet chromes are useful in blending with organic pigments such as organic reds and maroons. Blending scarlet chromes with quinacridone reds produces a range of colours which can be used in full shade as cheaper pigments with good quality.

Similarly to the toluidine red pigments, monoazo yellow pigments can be prepared. These were introduced under the trade-name of Hansa yellows, the most popular being a compound manufactured by coupling 2-nitro-*p*-toluidine with acetoacetanilide.

Hansa yellows, now known as arylamide yellows (7), have largely been superseded by other pigments in other industries but still find favour in the paint industry for decorative coating use. Like the toluidine red pigments the yellow arylamides have good durability in mass tone but have poor durability as tints. They tend to bleed in aromatic solvents and are only used in paints which contain weak solvents. They are lead free and replace lead chromes in full shades where lead-based pigments are restricted by legislation or for other reasons.

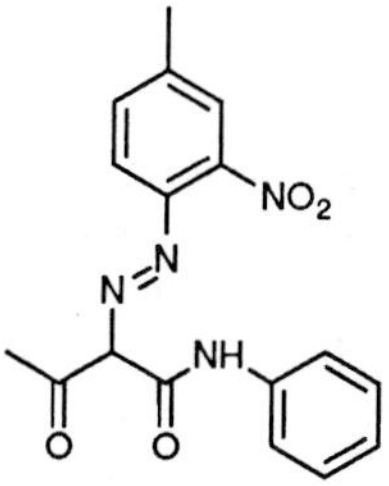

NO_2

N=N

NH

O O

(7)

Diazo yellow pigments which were known as benzidine yellows, now known as diarylide yellows (8), have improved tint strength, better solvent resistance, and give good opacity compared with monoazo yellows. Unfortunately, their light fastness is not so good as that of monoazo pigments. They are used extensively in the USA, especially in printing inks. The diamine used in the diazotisation is dichlorobenzidine with modified acetoacetanilides acting as the coupling reagents.

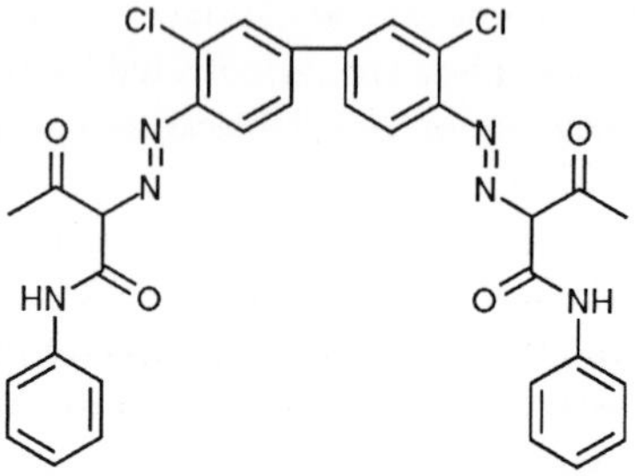

(8)

There is also a range of organic pigments known as Vat yellows (9) which have good light fastness in mass tone and as tints. They are based on polycyclic aromatic compounds such as flavanthrone and anthrapyrimidine but because of the high cost of production find use in a limited range of paint products, especially for the automotive industry.

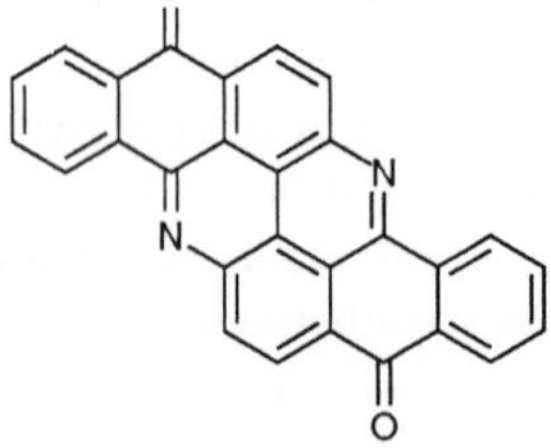

(9)

Black Pigments. The black pigments used in the protective coatings industry are of two types. Black iron oxide and carbon black. Black iron oxide, $Fe_2O_3.FeO$, is available as a mineral but is also made synthetically. It is relatively cheap, inert, with good light fastness and chemical resistance.

For finishes, to give good colour, black iron oxide is rarely used, except perhaps to tint white pigments to grey colours. It finds use in primers and build coats.

Carbon black is made in large quantities for the rubber industry. In the paint industry, it gives a superior black colour. It is made by heating oil, gas, or acetylene at high temperatures with limited air to give the carbon pigment.

Carbon black is a pigment with a very small particle size (sub-micron). The smaller the particle size the better the colour obtained. Carbon blacks cause severe problems due to their small particle size which gives a very large surface areas. The dispersibility of the pigment can be difficult. This can lead to problems of poor gloss and poor colour. Carbon black is also poor in giving stable paints with air drying media, and when used with other pigments such as titanium dioxide can flocculate changing the colour on storage.

Vegetable blacks, which were made by burning vegetable oils in a limited supply of air, have a larger particle size than carbon blacks but have poorer colour in that they appear dull and less brilliant than a good carbon black when formulated in gloss paints. They have limited use in tinting with other coloured pigments.

Pigmentation

A paint film is applied to an area at a required film thickness. It is the optimum thickness that controls durability and performance, and it is therefore the volume of the paint film which is important. It follows that the relationship of pigment to binder in a paint film must be a volume relationship and if we are to understand the effects of pigments in a coating this volume relationship must be appreciated.

Pigments, being powders, are not easy to measure as absolute volumes, it is far easier to weigh a powder in a laboratory or factory, so the first physical property of any pigment which should be determined is its relative density. From this, the volume formulation of a film can be calculated as a weight formulation for manufacturing purposes.

Pigments with high relative densities, for example red lead and barytes, are difficult to formulate as paints if high volume loadings are required. They give coatings with high densities which tend to settle, are difficult to handle, and heavy to apply. Pigments with low relative densities are far easier to use, especially at high volume loadings in films.

The volume relationship of pigment to binder in a film can be given by the simple, yet extremely useful, concept 'pigment volume concentration' (p.v.c.) normally expressed as a percentage:

$$\text{p.v.c.} = \frac{\text{Volume of Pigment}}{\text{Volume of Pigment} + \text{Volume of Binder}} \times 100 \qquad (1)$$

Obviously at 100% p.v.c. the 'film' will consist entirely of pigment, whilst at 0% p.v.c. the film is devoid of pigment. In between these two extremes a range of films can be produced with varying values of p.v.c. The films have different properties depending upon the value of the p.v.c.

The dispersion of a pigment in the medium or binder is a complex operation but the aim, especially with coloured pigments, is to break down the agglomerated pigment to near its original particle size and thoroughly 'wet out' the pigment with the dispersing polymer. This is necessary to achieve optimum properties of opacity, durability, colour, and particularly to obtain 'value for money' as coloured pigments are generally very expensive. When dispersing the pigment, its surface has to be stabilised by adsorption of polymer to produce colloidal stability which prevents flocculation of the pigment particles or at least reduces flocculation to a minimum when the paint is made and stored.

Pigments which have high surface areas, *i.e.* small particle size, will require larger amounts of polymer to stabilise the colloid compared with pigments which have large particle size. This can be demonstrated by a simple measurement of oil absorption.

In this measurement a known weight of pigment is wetted with linseed oil by slowly adding the oil, drop-wise, whilst the pigment is rubbed with a spatula on a glass slab. The end point of the titration is the point at which the pigment particles are held together to form a mass rather than a crumbly solid. The amount of oil required to reach this point is measured and the oil absorption is expressed as the weight of oil absorbed by 100 g of pigment.

The end point of the oil absorption measurement could represent the point where the pigment particles are fully wetted out and the particles are packed to an optimum state. Voids present between the particles are filled with oil, there is no excess and there is no dry pigment present. This point has become known as the 'critical pigment volume concentration' (c.p.v.c.).

Many workers have demonstrated that paint films formulated to the c.p.v.c. give optimum values for certain properties. These are mainly related to durability of build coats or primers. It can be imagined that at this point on the p.v.c. scale there is minimum free binder present in the film, and this will give maximum resistance to the transmission of ions and molecules through the film. Hence, at the c.p.v.c. it would be expected that there would be maximum resistance to corrosion, and maximum resistance to blistering of the film as it should have least expansion due to water uptake by the polymer. It could also be considered that at the c.p.v.c. the cost of the durable dry film could be at its lowest if the pigments being used were cheaper than the medium per volume. With certain paints, p.v.c.s much higher than the c.p.v.c. are used but these are mainly decorative matt wall paints.

The concepts of oil absorption and c.p.v.c. are quite simple. The oil absorption value, as determined using linseed oil, cannot be translated to other polymers as each polymer will have different dispersion properties. In most paints, mixtures of pigments and extenders are used and, as they are present in differing sized and shaped particles, the way they disperse and pack in real films can be quite different from the way they behave when oil absorption measurements are made. Nevertheless, oil absorption figures can give chemists a parameter from which they can work in formulating coatings. High oil absorption values give low values for the c.p.v.c., whilst low oil absorption values give high values for c.p.v.c.

Especially with barrier coatings there is a pigment volume concentration which should not be exceeded if maximum benefit of durability is to be achieved with any pigment blend. The value of this p.v.c., which can be named the c.p.v.c. for the film, in many cases has to be determined by experiment. Paint films are prepared with varying p.v.c.s, below and above the expected c.p.v.c., and durability studies made to determine the pigment to binder ratio which is appropriate for the environmental conditions which the paint film has to withstand.

For anti-corrosive primers, although the concept of the c.p.v.c. is important in obtaining the desired properties of impermeability, blister resistance, adhesion, and film strength, there are other considerations which have to be taken into account. They are the correct volume of the anti-corrosive pigment to give good protection and long life, the relationship of volume of the anti-corrosive pigment to other pigments, and the amount of colouring pigment necessary to give colour and opacity. The type of barrier effect pigments and, as already emphasised, the total cost of the primer is crucial as it must be competitive in the market place.

To establish the most cost-effective formulation of anti-corrosive primers, experimental studies are carried out most effectively using accelerated durability tests such as salt spray tests, accelerated weathering studies, immersion tests in solvents and solutions, adhesion and mechanical strength measurements, together with overcoating examinations. All of these experiments are time consuming and expensive to perform. It is prudent, therefore, if it is at all possible, to have a statistical approach to formulations to reduce the number of experiments to obtain the most effective formulation.

The various tests which are performed to obtain the technical information should be recognised standard tests, preferably to International Standards. This is to ensure that the tests are reproducible and the information is accepted by industry, though their correspondence with real service conditions is often questioned.

The concept of c.p.v.c. is important for anti-corrosive and barrier coatings but for finishing paints, particularly those with a high gloss, pigment properties are of greater importance; notably of coloured, white, and black pigments which are used to colour and opacify a film. In the case of a typical pigment, such as titanium dioxide, these properties are controlled by the refractive index and the particle size of the pigment.

The effect of refractive index can be shown by the use of a form of Equation 2 derived by Fresnel:

$$\text{Reflectivity of a coating} = \frac{(\eta_p - \eta_r)^2}{(\eta_p + \eta_r)^2} \qquad (2)$$

where η_p is the refractive index of the pigment, and η_r is the refractive index of the medium.

This leads to the conclusion that the higher the difference between the refractive index of the pigment and the medium the greater the reflectivity of the film. Most polymers used in the paint industry have a refractive index of about 1.5. Titanium dioxide at 2.7 has the highest refractive index of any white pigment. Zinc oxide has a value of 2.0, whilst that of white extenders, such as barium sulfate, is around 1.6.

Using the Fresnel equation (2) if the relative opacity of titanium dioxide is taken as 100, zinc oxide will give a relative opacity of 25, and barium sulfate 0.2. In other words, titanium dioxide will give opaque glossy films, whilst barium sulfate will only give translucent glossy films. However, at very high p.v.c.s, greater than the c.p.v.c., air is present in the film and as air has a refractive index much lower than that of the binder, extenders can give opaque matt films.

It is interesting to note that the element carbon gives the best black pigment, and also could give an excellent white pigment in the crystal form of diamond which has a very high refractive index. Unfortunately, diamond is rare and its physical properties of hardness and particle size are not conducive to manufacture of a pigment.

The particle size of a pigment controls the light scattering effect and this affects its opacifying property. For titanium dioxide, the primary particle size of 0.24 μm is approximately half the wavelength of light, which is optimum for light scattering by diffraction. It is unlikely that a dispersion of titanium dioxide ever reaches the state where all of the agglomerates are broken down to the primary particle size, but this demonstrates the importance of having good dispersion at the correct pigment volume concentration in order to achieve good opacity.

Finishing paints which are formulated to have high gloss have p.v.c.s which are much lower than the c.p.v.c. Ideally, a clear film will give maximum gloss, pigmentation lowers gloss, but for most practical purposes the gloss of pigmented films will remain constant with increasing p.v.c. until the volume of pigment reaches a point where particles of pigment disturb the surface of the film and cause micro-roughness. With solvent soluble polymers such as alkyds, urethanes, epoxies, and others, it can be shown that a clear layer of unpigmented polymer is present at the surface of the film at low p.v.c.s. It is the disturbance of this clear layer by the increasing p.v.c. which lowers the gloss of finishing coatings, but this is gradual until the surface becomes rough due to the presence of pigment. Increasing the p.v.c. further lowers the gloss which becomes minimum at the c.p.v.c.

A compromise has to be made with high gloss films in the relationship between opacity and gloss. Other factors which also control this are the degree of dispersion of the pigment, and the film thickness of the applied film.

Once the pigment is dispersed to the degree required to give opacity and gloss, the dispersion must be stabilised to avoid flocculation of the pigment which could reduce the opacity and gloss.

Typical values of p.v.c. for gloss paints are 10–25%, whilst those of barrier coats and anti-corrosive primers are much higher, nearer 30%.

With the exception of white and black, very few standard coloured paints are produced from a single pigment. The number of coloured paints which can be prepared is almost limitless. Even if a colour is selected which can be made from a single pigment, it is difficult for the pigment manufacturer to produce a consistent colour, in every respect, from one batch of pigment to another. It is also difficult for paint manufacturers to produce, consistently, the same colour from one batch of paint to another. In some part of the process additions have to be made to correct the colour. Most colours are achieved with mixtures of pigments.

Let us suppose that a pale colour is required. It can be produced from titanium dioxide with other pigments, depending upon the colour. In the manufacture of the paint several possibilities are available. The titanium dioxide can be dispersed, the paint finished by additions of all of the other constituents, driers, solvents, *etc.*, and then the colour can be produced by the addition of a dispersion of the colouring pigment. At this stage it is possible to add a number of dispersions of different colours to achieve the colour. This would be an efficient way of producing coloured paint, as the primary dispersion of the base colour, white, could be made on a large scale, and with the separate dispersion of coloured pigment a very large colour range could be produced even in relatively small quantities.

This is the principle of 'in store' production of retail colour paints, which is very popular in the USA and less so in Europe. The principle can be extended further by the use of universal colour dispersions which are capable of colouring white bases formulated from different polymers. For example, it is possible to have white alkyd, urethane, and epoxy bases all being tinted to shade using the same coloured dispersions. The technology depends upon the coloured pigments being dispersed in a medium which is compatible with all of the polymers used in the white bases. It is also possible to include water-based white bases in the universal colouring systems.

These tinting or colouring systems can apply to any sized batch of coloured paint, be it a thousand litres or one litre. They can be used in factory made colour paint processes or at point of sale processes. In that way, rapid response can be given to requests for coloured paint in any reasonable quantity; stocks of finished coloured paints can be reduced to a minimum; and the number of manufacturing processes involving dispersion can be minimised. These objectives are made possible by the use of computer controlled techniques for the measurement of colour and for the prediction of colour from known standards. There are many variations of the way in which coloured paint can be made. The industry being served commands the techniques used to produce the end product in the most economical manner. Research and development proceeds to develop new pigments with improved properties, to develop new techniques of colour measurement and dispensing so that the demand for colour and colour fashion is satisfied.

SOLVENTS

The majority of resins traditionally used as media for paints are solids, or highly viscous materials, which have to be diluted with a liquid to lower their viscosity so that they can be manufactured into paint and the paints then applied to substrates.

The liquid can act as a true solvent for the resin, or as in the case of water can act as a diluent in water-based emulsion paints.

All solvents for paint resins are low molecular weight organic compounds, most of which can be classified as aliphatic or aromatic hydrocarbons and oxygenated organic compounds.

The most important property of a solvent is to completely dissolve the resin to give a clear solution which can be used to disperse the other ingredients of the paint formulation. The pigment dispersion is corrected for viscosity by the further addition of solvent so that the paint can be applied by brush, spray, or other techniques. After application of the paint the solvent evaporates from the film to give a surface which is smooth, uniform, of the correct gloss, and free from imperfections.

The role of solvents is very significant in that they control many factors associated with the properties of paint films. The physical and chemical characteristics of solvents which control paint properties are as follows:

Solvency. A true solvent should dissolve a resin completely to give a clear solution over a wide range of concentrations, and, if the solution is applied as a film, the solvent should evaporate to give a transparent film free from imperfections.

The molecular forces which affect solvency are dispersive forces (London forces), polar forces, and hydrogen bonding. These forces also control solvent properties such as boiling point, surface tension, and latent heat. The solubility parameter, which is a function of the cohesive density of liquid, is also an important factor in the appreciation of the solvent power of solvents. For solvents and polymers to be miscible they should have very similar solubility parameters. The effect of polarity of the solvent and polymer is very important in the relationship of their solubility parameters.

In practical paints it is usual that more than one solvent is used in a formulation. In many cases, mixtures of a true solvent and a diluent are used as the 'solvent'. A diluent being a liquid which is not a true solvent. The reasons why mixtures of solvents are used can include the control of evaporation, *etc.*, control of flash point, cost, and also to reduce health and safety risks. Nevertheless, whether a paint solvent is a single liquid or a mixture of liquids, it must have high solvent power over the whole time it is present in the paint film. Premature evaporation of the 'stronger' components in a solvent mix would lead to precipitation of polymer.

Addition of too much diluent may produce a θ-solvent whose mixtures with polymer show increased viscosity (*i.e.* polymer–polymer interactions predominate over polymer–solvent interactions and the chains are pictured as coiling up like the Greek letter). At still higher levels, the polymer will be precipitated.

The phenomenon of latent solvency is of considerable importance in the formulation of solvent-based paints. For example, mixtures of ethanol and toluene or butanol and xylene, none of which is a satisfactory solvent in its own right, show excellent solvency for certain polymers.

Viscosity. Organic solvents are low viscosity Newtonian liquids. Solvency, interactions, and concentrations being equal, a lower viscosity solvent will produce a lower viscosity polymer solution.

Flash Point. As solvents are low molecular weight organic compounds there is a fire and explosion risk when they are used in the manufacture, storage, transport, and application of paint. The flash point, which is the lowest temperature at which the vapour of the liquid can be ignited in air under defined conditions, has to be taken into account in the formulation of paints. There are strict regulations governing the transport and storage of paint, and the flash points of paints have to be determined by specific methods to comply with the legislation. With a mixture of solvents it cannot be assumed that the flash point will be that of the solvent with the lowest flash point. Mixtures of xylene and butanol, for example, have flash points which are lower than those of either component.

Evaporation Rate. A solvent should evaporate from a paint film at a rate suitable for the method of application. For example, the evaporation rate in a brushing paint should be such that the applied film remains fluid for a time so that the 'wet edge' remains open, and areas of wet film can be easily joined without showing signs of poor flow, *i.e.* brush marking.

For spraying paints it is important that one solvent evaporates very quickly and the wet film increases in viscosity to avoid excessive flow and runs (*c.f.* Chapter 4). Remaining solvent in the film evaporates more slowly and thus allows the wet film to smooth to a fully coalesced, pinhole free film.

The rate of evaporation of solvent from a paint film can vary widely unless the atmospheric conditions of application and drying/curing are controlled. For factory applied paint this generally is not a problem, but for protective coatings which have to be applied during winter and summer, in dry and humid conditions, the choice of solvents and solvent mixtures can be very important if problems of poorly applied and poorly cured films are to be avoided.

The retention of solvents in films is also a factor which has to be considered, as it can lead to blistering of immersed films due to osmotic effects of, for example, retained oxygenated solvents which are soluble in water. It is largely governed by the vitrification of the polymers present (*c.f.* Chapter 5).

Two classes of coatings constituents which may be loosely regarded as solvents are reactive diluents and plasticisers.

Reactive diluents are mobile, reactive low molecular weight materials used to reduce the viscosity of compositions. They are subsequently combined with the cured material rather than being lost to the atmosphere (*c.f.* Chapter 7). In reality, molecules small enough to be mobile will have appreciable volatility, and their satisfactory design and use depends on maximising reactivity and minimising evaporation rate.

Plasticisers are inert materials of very low vapour pressure which are intended to remain in a coating film throughout its life to modify its mechanical properties. Typical materials are phthalate and phosphate esters and certain polymers. In practice, the lower molecular weight types migrate or slowly evaporate with shrinkage and embrittlement, and it is generally better to modify the film-forming resin to provide the required properties ('internal plasticisation').

Toxicity. Most solvents must be treated as toxic materials, and legislation is in force in all industrialised countries to control their use. They can enter the body by inhalation, by skin contact, and by ingestion. Measures have to be taken to reduce risks to humans to a minimum. These can be the use of simple protective clothing, such as gloves for application by hand, or complete air-fed suits with visors for spray application in enclosed areas which are also ventilated to prevent build up of vapour.

Toxicity studies using animals and micro-organisms provide data to assist in the control of toxic hazards. Official industrial hygienists in various countries publish lists of substances with atmospheric concentrations to which workers can be exposed safely. The best known values are the Threshold Limit Values (TLVs) produced in the United States. They are weighted averages which take into account the time in which a person can work in an atmosphere containing the solvent. The TLV is expressed as parts per million in their air, and should not be exceeded. In the UK, a similar system uses Occupational Exposure Limits (OELs). TLVs or OELs can be taken as a working guide to the toxicity of paint solvents but it is important to be alert to changes. In recent years, for example, the popular solvent 2-ethoxyethanol, which had been considered a relatively benign material, showed severe effects in animal tests. Its TLV was dramatically lowered and in many industries it was replaced by the more expensive methoxypropanol.

Environmental Properties. The association of solvents with industrial smog in places such as California, the depletion of the ozone layer, and the introduction of legislative controls was discussed in Chapter 2. Some products, for example chlorinated solvents, will probably have to be phased out within a relatively short time. It is quite apparent that all organic solvents are being considered an environmental problem which has to be greatly reduced. This is forcing the development of water-based and solvent free coatings to be a major priority in the coatings industry.

Odour. The odour of paint solvents has always been a contentious subject, and efforts are made to reduce the nuisance to a minimum. Several attempts have been made by the oil industry to produce solvents, such as aliphatic hydrocarbons, with very low odour to overcome the problem.

Solvents are manufactured by many processes but all are derived from petroleum nowadays. Hydrocarbon solvents are fractions obtained at the same time as fuels, whilst other solvents are manufactured substances which may also have uses as raw materials for the synthesis of organic chemicals, plastics, polymers, and pharmaceuticals.

Typical aliphatic hydrocarbon solvents are hexane, cyclohexane, 'white spirit' of various distillation ranges, and special fractions manufactured for particular uses.

Aromatic solvents are xylene mixtures ('xylole'), toluene, and various trimethyl benzenes. Benzene itself is not allowed as a paint solvent because of health hazards. Tetrahydronaphthalene ('tetralin') finds limited use.

The oxygenated solvents are:

Alcohols—methanol, ethanol, and isomers of propanol, butanol, and higher alcohols.

Ketones—acetone, methylethyl ketone, methylisobutyl ketone, diisobutyl ketone, and cyclohexanone are typical of the class.

Esters—ethyl acetate, butyl acetate, and mixed acetates.

Glycol ethers and derivatives—ethoxyethanol, butoxyethanol, methoxypropanol, and their acetates.

The volume of solvents used by the paint industry is relatively small compared with the total volume of organic liquids which are produced mainly as fuels. Paint solvents have only two uses: to allow paint to be manufactured, and to allow paint to be applied. Most conventional paint solvents are lost to the atmosphere without change or are burnt as waste products. This causes concern as it is wasteful and an environmental hazard.

The need to minimise the use of paint solvents has been the driving force in the development of alternative water-based coatings and solvent free coatings, and this will become more urgent as legislation is tightened.

BIBLIOGRAPHY

'Pigment Handbook, Vol. 1, Properties and Economics', 2nd Edn., ed. P. A. Lewis, John Wiley and Sons, Chichester, 1988.

R. Lambourne, 'Paint and Surface Coatings, Theory and Practice', Ellis Horwood, Chichester, 1987.

H. F. Payne, 'Organic Coating Technology', Vol. 2, John Wiley and Sons, Chichester, 1967.

T. H. Durrans, 'Solvents', 8th Edn., Chapman and Hall, London, 1971.

Chapter 10

Surface Coatings—The Future

A. R. MARRION

There are as many predictions of the future as people offering them, and each is as valid as the other, until the future arrives and proves it wrong.

It would be unwise to dwell over long on the future of the Coatings Industry or to attempt to be unduly specific. The industry is constantly evolving to meet new challenges and no doubt will continue to do so. Like every aspect of human endeavour, it will be conditioned by the constraints and opportunities encountered during its progress.

Amongst foreseeable constraints, the development of materials with satisfactory appearance and resistance to the elements is frequently seen as liable to eliminate surface coatings from various markets. Self-coloured plastics and weathering steels are two examples. However, capricious human nature can be relied upon to demand at least one colour change during the lifetime of many such artefacts and, in any event, the emergence of new markets will provide opportunities to compensate for any losses.

The gathering pace of environmental and health awareness and legislation will no doubt drive coatings technology still further along the lines suggested in earlier chapters. Organic solvents will be shunned except in certain specialised situations, and an increasing variety of solventless coatings will reach the required level of technical excellence.

In many cases the ability to make well 'tailored' polymers—telechelics, stars, or dendrimers—of specified molecular weight distribution will be decisive in controlling the rheology and other physical characteristics upon which such technologies will depend.

Water-based coatings will continue their rapid growth, with polymer and additive developments allowing further reductions in their organic solvent contents. Powder coatings will form an increasing proportion of stoving industrial coatings, though considerable improvements in their cosmetic properties will be needed if they are to penetrate the automotive top coats market, for example. High solids paints will be refined to higher solids, and offer solutions to solvent emission problems in the medium term. However, in most areas legislation will eventually demand essentially '100% solids' coatings and it is unlikely that incremental improvements will always provide suitable technologies. The dramatic expansion in demand for radiation cured coatings is evidence of the success of one solvent free innovation.

The design of the new generations of polymers and coatings will be greatly facilitated by computer modelling of molecules and 'expert' formulation systems linked to statistical experimentation.

The frequency with which chemical substances are found to be harmful to the human body will also continue to encourage innovation. We may expect to see a succession of candidates for replacement of curing agents which contain isocyanates, or release formaldehyde, and for 'effect' pigments, for example the anti-corrosive chromates. The ideal, and increasingly necessary situation, is one in which all hazardous materials are eliminated from coatings formulations. The technical cost may be high, but penalties for failure are higher, including the risk of harming users, members of the public, or the environment.

Concern about waste disposal will assume enormous importance. Coatings manufacture is generally not detrimental to the environment since a high proportion of the raw materials entering a factory subsequently leave in the form of product. The same cannot be said for some coatings raw materials where operations such as washing involve significant waste streams at present. Such processes are likely to be subject to review with consequent impact on the economics of paint formulation.

The use of coatings materials often does create waste in the form of empty cans and water or solvent contaminated by cleaning spray guns or brushes. Legislation in Europe, the United States, and other parts of the world imposes a lifetime duty upon manufacturers with regard to their products; coatings suppliers will therefore be responsible for safe disposal of contaminated cans and paint residues and will have to introduce significant changes in their work patterns. Recycling of containers and materials will become mandatory. Waste disposal is already carefully controlled and contamination of land or water courses with toxic materials is to be avoided at all costs.

The imminent exhaustion of raw material supplies has been predicted for many years, though new sources, or technologies using different materials have usually been forthcoming. In particular, the world's petroleum resources are finite, and ready access to a range of organic chemicals may soon be lost.

Apart from the obvious incentive to conserve materials through recycling and use of the most efficient application methods, the potential depletion of petroleum resources has stimulated interest in polymers derived from renewable resources. Extended use of, or reversion to, some of the traditional natural materials (drying oils, cellulose derivatives, natural resins), or the introduction of new plant products with useful properties (such as vernonia oil, *c.f.* Chapter 7) is anticipated. More significantly, biotechnology will ultimately provide the means of generating highly specific materials, such as the 'well tailored' polymers mentioned above, from inexpensive biomass. It will also enable them to be manipulated in appropriate ways—water-based materials rendered hydrophobic after application, for example.

On the other hand, since the growing environmental mood is against squandering oil and coal as fuels, there may after all be no shortage of organic feed stock for many centuries.

The most satisfactory view of the future is that the Industry will continue to grow and provide an important service to a widening range of surfaces with increasingly well designed and environmentally benign products. Still higher levels of performance will be offered by new chemistry and technology.

The increasing use of fluoropolymers is but one example of new chemistry offering remarkable new opportunities—a level of soiling and weathering resistance which allows the frequency of repainting to be greatly reduced.

The use of inorganic coatings, such as metals and ceramics, for high temperature service, and/or chemical resistance, or long life in inhospitable places represents a dramatic departure from conventional organic polymer-based coatings. Conversely, heat resisting organic polymers are replacing vitreous enamel on domestic cookers where they offer better mechanical properties and a wider range of colours.

There will be many other examples in coming years, but much will depend on the imagination and serendipity of the coatings technologist.

Glossary of Trivial Names and Technical Terms

ACETONE Propanone.

ACRYLAMIDE Propenamide.

ACRYLATE Propenoate.

ACRYLIC A polymer obtained by polyaddition of acrylates, methacrylates, and similar monomers.

ACRYLOYL Propenoyl.

ALKYD A class of coatings polymers; polyester bearing pendant fatty acid residues, to provide auto-oxidative drying or plasticity.

ALLYL Prop-2-enyl.

AMBIENT CURE A curing reaction which takes place at the temperature prevailing on site.

ANAPHORETIC See anodic.

ANODIC Anaphoretic. Related or transported to the anode in a cell. Electropaints containing polymeric anions which coat the anode or positively charged electrode.

AUTO-OXIDATION The interaction of organic materials with singlet oxygen to produce free radicals resulting in crosslinking and/or degradation.

AZIRIDINE Ethyleneimine, azacyclopropane.

AZLACTONE Oxazoline-5-one; 2-aza-4-oxo-5-oxacyclopent-1-ene.

BAKING CURE See Stoving cure.

BINDER Vehicle, medium. Polymeric or polymerisable components which provide the cohesion of a coating composition.

BISPHENOL A 2,2-Bis(4-hydroxyphenyl)propane. 4,4′-Isopropylidene diphenol.

BUILD COAT See Surfacers.

BUTYL ACETATE Butyl ethanoate.

CATAPHORETIC See Cathodic.

CATHODIC Cataphoretic. Related or transported to the cathode in a cell. Electropaints containing polymeric cations which coat the cathode, or negatively charged electrode.

CHLOR-RUBBER Chlororubber, chlorinated rubber. A polymer of complex constitution made by chlorinating natural rubber.
CHROMA The intensity of 'saturation' of a particular hue when compared with a neutral grey. Spectral colours have the highest chromas.
CONVERTIBLE A coating which undergoes chemical change after application.
CONSISTENCY A general term for the property of a material by which it resists permanent change of shape.
CROSSLINK That part of a polymer network which links chains, often formed previously, to one another.
CROSSLINKING AGENT A substance which can bring about crosslinking of a reactive binder.
CURE The process by which a freshly applied coating becomes intractable. Often synonymous with crosslink.
CURING AGENT A substance which can bring about curing of a reactive binder, by co-reaction or catalysis.
CYCLIC ANHYDRIDE 1,3-Dioxo-2-oxacycloalkane derivatives often five membered, therefore -cyclopentane.
CYCLIC CARBONATE 1,3-Dioxa-2-oxocycloalkane derivatives often five membered, therefore -cyclopentane.
DIALYSIS A process for separating low molecular weight species from polymer dispersions by diffusion through a semi-permeable membrane.
DIISOBUTYLKETONE 2,6-Dimethylheptan-4-one.
DURABILITY In coating terminology ≡ weatherability, *i.e.* degree of resistance to degradation in the environment.
ELECTROPAINT Electrophoretic paint; electrodeposition paint. A coating system in which an ionised polymer is made to deposit on the electrically charged workpiece immersed in a dispersion of the polymer and other components.
ETCH PRIMER See Wash primer.
ETHYL ACETATE Ethyl ethanoate.
EPICHLOROHYDRIN 1-Chloro-2,3-epoxypropane.
EPOXIDE See Epoxy.
EPOXY Epoxide. A polymer or reactive species containing epoxy or oxirane groups, $\overset{O}{\diagup\diagdown}$ C—C, or derived from such materials.
EXTENDER A powdered material, usually a mineral, used in coating compositions to impart various properties, but principally to reduce cost.
FATTY ACID A long chain monocarboxylic acid derived from fat or

oil. Often containing auto-oxidisable allylic groupings.

FILLER See extender.

FILM FORMER In coatings technology, the part of a composition which imparts the ability to be spread on a surface as a thin, adherent film which will subsequently solidify. Usually a polymer.

FORCED CURE A curing reaction accelerated by moderate heating (usually between room temperature and about 100 °C).

FORMALDEHYDE Methanal.

FUMARATE *trans*-Butendioate.

FUNCTIONAL MONOMER A monomer containing an additional reactive group, available for reaction after polymer formation.

GALVANISED STEEL Steel coated with zinc to give sacrificial corrosion protection.

GEL COAT In glass reinforced polyester technology, a pigmented composition applied to the surface of a mould before applying lay-up resin. It becomes the coating of the artefact on removal from the mould.

GLASS TRANSITION Vitrification. A phase change in a polymer from rubbery to glassy, over a specified temperature range, the T_g.

GLYCIDYL 2,3-Epoxypropyl.

ITACONATE 2-Methylidenebutanedioate.

MALEATE *cis*-Butenedioate.

MALONATE Propanedioate.

MEDIUM See Binder.

MELAMINE *sym*-Triaminotriazine.

MELAMINE RESINS Condensation products of melamine with formaldehyde and optionally alcohols.

METHACRYLATE 2-Methylpropenoate.

METHYLISOBUTYLKETONE 4-Methylpentan-2-one.

METHYLETHYLKETONE Butanone.

METHYLOL Hydroxymethyl.

β-NAPHTHOL 2-Hydroxynaphthalene.

NETWORK A crosslinked polymer. A molecule of essentially infinite molecular weight.

NEWTONIAN FLUID A fluid for which the shear stress is proportional to the shear rate.

NITRO-TOLUIDINE Nitroaminomethylbenzene, various isomers.

NON-NEWTONIAN FLUID A fluid for which the proportionality between shear stress and shear rate is not constant with shear rate.

OIL LENGTH A measure of the amount of fatty acid in a polyester (calculated as triglyceride).

OLEFIN Alkene.

OLEORESINOUS MEDIA Coating compositions compounded from oils and natural resins, usually cooked together at high temperature.
OLIGOMER A moderate molecular weight species formed from a few monomer residues. Often used in the same sense as 'resin' in the coatings industry.
OXAZOLINE 1-Aza-3-oxacyclopent-1-ene derivatives.
PHENOL Hydroxybenzene.
PHTHALIC ACIDS Isomers of benzendicarboxylic acid; 1,2- = orthophthalic; 1,3- = isophthalic; 1,4- = terephthalic.
PIGMENT A powdered material used in coating compositions to impart various properties, most notably colour.
POLYCAPROLACTONE Poly-1,6-hexanoide, poly[oxy(1-oxohexamethylene)].
POLYESTER A polymer containing ester linkages –CO–O– in its backbone.
POLYETHER A polymer containing ether linkages –O–.
POLYMER A high molecular weight species formed from many monomer residues.
POLY(METHYL METHACRYLATE) Poly(methyl 2-methylpropenoate), poly[1-(methyoxycarbonyl)-1-methylethylene].
POLY(STYRENE) Polyphenylethene.
POLYUREA A polymer containing urea linkages, –NH–CO–NH–.
POLYURETHANE A polymer containing, or a coating system potentially containing, urethane or carbamate linkages, –O–CO–NH–. Often used loosely to include polyureas.
POLY(VINYL ACETATE) Poly(ethanoyloxy-ethene).
POLY(VINYL ALCOHOL) Poly(hydroxyethene). Hydrolysed poly(vinyl acetate) containing variable levels of residual ester.
POLY(VINYL BUTYRAL) Butyral of poly(vinyl alcohol). A polymer containing dioxacyclohexane residues. Poly[2-propyl-1,3-dioxon-4,6-diyl)methylene].
POLY(VINYL FLUORIDE) Poly(fluoroethene).
POLY(VINYLIDENE FLUORIDE) Poly (1,1-difluoroethene).
POWDER COATING A coating applied as a dry powder which is fused to a coherent film by heating.
PRIMER A coating applied to a bare substrate to prepare it for subsequent coats, *e.g.* to seal pores and promote adhesion.
REACTIVE BINDER A coating polymer bearing reactive groups for network formation.
REACTIVE DILUENT An additive which behaves like a solvent in reducing viscosity, but combines with coating polymers as they cure.
RED LEAD Lead tetroxide, 'suboxide', minium Pb_3O_4.

RESIN A film forming material, usually polymeric or oligomeric.
ROSIN A natural resin derived from pine trees.
STOVING CURE A curing reaction which takes place at deliberately elevated temperature (usually above about 100 °C).
SURFACER A thick coating designed to cover imperfections in a substrate prior to the application of a top coat. It may be formulated to permit sanding.
TELECHELIC A well-defined end-functional polymer or oligomer.
TETRAHYDROFURAN Oxacyclopentane.
THERMOPLASTIC A polymeric system which can revert to the fluid state on heating. A non-convertible system.
THERMOSET A polymeric system which can undergo (or has undergone) a chemical crosslinking process, usually on heating.
THIXOTROPY A slow recovery of consistency lost by shearing when the sample stands undisturbed.
TOLUENE Methylbenzene (occasionally rendered toluole or toluol in the coatings industry).
TOLUOLE See Toluene.
TOP COAT The last layer in a coating scheme, providing the colour, gloss, and weathering properties required.
ULTRAFILTRATION A process akin to dialysis, but using moderate hydrostatic pressure and high fluid velocity to assist separation.
UNDERCOAT A coating designed to prepare a substrate for the top coat, *e.g.* to provide a smooth surface of a colour which will not show through the top coat.
UNSATURATED POLYESTER A polyester containing unsaturated residues (derived from maleic anhydride, for example) to permit crosslinking in the presence of free radical source and optionally other olefinic materials.
VEHICLE See Binder.
VISCOSITY The resistance of a material to deformation. The shear stress divided by shear rate.
VITRIFICATION See Glass transition.
WASH PRIMER A thin coating used for preservation of steel components in storage. It usually contains phosphoric acid to passivate the metal surface.
XYLENE Dimethylbenzene (three isomers are recognised).
XYLOLE, XYLOL Words used in the coatings industry to describe the commercial mixture of xylenes and ethyl benzene used as a solvent. The former is preferable.

Subject Index